BLACK COUNTRY
BREWERIES

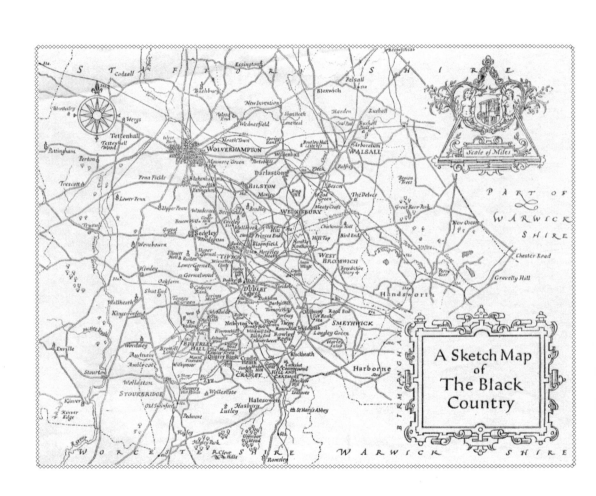

A Sketch Map of The Black Country

BLACK COUNTRY BREWERIES

JOSEPH McKENNA

The
History
Press

First published 2005

Reprinted in 2010 by
The History Press
The Mill, Brimscombe Port,
Stroud, Gloucestershire, GL5 2QG
www.thehistorypress.co.uk

Reprinted 2011

British Library Cataloguing in Publication Data.
A catalogue record for this book is available from the British Library.

ISBN 978 0 7524 3722 4

Typesetting and origination by Tempus Publishing Limited.
Printed in Great Britain.

CONTENTS

ACKNOWLEDGEMENTS

My grateful thanks to the staff of Brierley Hill, Dudley, Walsall and Wolverhampton Public Libraries. My thanks also to Paul Taylor and Andy 'Bruce' Willis for scanning. Lastly to Chris Ash for his photographs and giving up his time to accompany me in all weathers in my search for the Black Country's lost breweries.

ONE

EARLY BREWING AND EARLY BREWERS

The Black Country lies within the West Midlands and straddles the old borders of Staffordshire and Worcestershire, touching upon north Warwickshire near Birmingham. It has no clearly defined boundaries and yet is a very real and distinct region in England. Its people speak an old form of English and brew the best mild in the country. Records regarding the brewing of ale in the area go back to 1468, when the Borough of Halesowen licensed and regulated five common brewers:

> One brew in the street between the Church and end of the town leading towards Cradeley, and in the High Street one brew, and from Laconstoon by the High Cross up to Cornebowe one brew, and from the said Cornbowe by the high Cross up to Laconstoon one brew, and in the street called Birmyngeham street one brew, under penalty of each one 6s 8d whereof 40d thereof shall be levied to the use of the lord, and 40d to the use of the Parish of Hales. And so in this manner that all brewers shall brew alternately from week to week, so that there may be five of new and five old.

From the fourteenth century Ale Conners, or tasters, were appointed to check the quality and measurement by which the ale was sold. By coincidence in 1477 one of the Ale Conners from Halesowen was one William Beare. In 1570 the sale of ale in the town was further regulated by the fixing of prices for its sale:

> All brewers that brew ale to sell shall brew good ale and wholesome for a man's body at 3d a gallon and 1d a gallon to be warranted by the ale taster. And that they do not sell by the cruse or cup but by the pewter pot of a lawful size in pain of 3s 4d.

Among the many ways the Church raised money for repairs to its buildings was the brewing and sale of ale known as Church Ale. In 1502 the Bishop of Coventry and Lichfield wrote to the Mayor of Walsall regarding the regulating of such events. Drinking, he said, should take place four times a year, and everyone attending should pay 1d. At Whitsuntide 1505, Halesowen parish church raised £3 13s 4d through such an event. Church Ale was banned during Oliver Cromwell's Protectorate. It enjoyed a brief revival under Charles II but, lacking the gusto of former days, fell out of use by 1700.

Its failure was due in part to competition from other sources, not only common brewers, but domestic brewers too. Probate Inventories, or lists of the contents of houses belonging to the recently dead, give details of both kinds. The inventory of John Wollaston, gentleman of Walsall who died in 1634, lists:

In the Brewhouse	
Item one greate mashing fattes [vat]	*4s 0d*
Item 2 greate yeelinge fattes [vat]	*6s 0d*
Item one grete clubb for swines mash	*3s 4d*
Item one iron fyre forke & 1 iron peale [pole?]	*6d*
Item one table board & one frame for the brewing	
fatt [vat] to stand upon	*3s 2d*

Abraham Powell, believed to be landlord of the *Old Chapel Inn*, Smethwick, died in 1737. His brew-house contained *'One Copper furnace & Brewing Vessels, two potts & a sauce pan'*. In his cellar were *'Four hogsheads six lesser barrel[s] & an Ale vat, one hogd, & half of Ale, bottles & bottle rack'*. His Great Parlour appears to have acted as his bar room. It contained *'Two Round tables & a screen, fourteen chairs & a Stafford Grate'*.

John Freeth, brewer and landlord of the *Red Cow* in Smethwick, also died in 1737. His Inventory, relating to his calling, includes the following:

In the Parlour one grate table and two little tables, 4 lethern cheers [chairs] and 2 others, two benches and a furm	*£0 19s 2d*
In the Seller 5 half hogshots and drink in them and benches As they stand on, 1 little old table 10 shilfs [shelves] 3 Duzzen of glass bottles, Earthen ware of all Sorts	*£1 6s 0d*
In the Brew house 1 furnish 1 Noiding trough 1 Table and frame, 1 Churn 8 tubs 1 Cheesepres and 4 shilfs, 1 pair of Ceepers, 1 pair of geals, 2 Sivs and other things there	*£2 1s 6d*

And if you were poor and forced to go into the workhouse, beer was still available, albeit 'small beer', a very weak beer. In Walsall 'The Provisions for the Workhouse, from the 12th to the 19th April 1733', lists:

For hops and barm, and for brewing	*£0 1s 1d*

In 1801, with a population exceeding 100,000, there was not a single common brewer in the Black Country. The first known was William Jones of Jones & Co., brewers of 7-9 Snow Hill, Wolverhampton. The company appears in the town Rate Book of 1802. Thereafter nothing more seems known of Jones or his brewery. Three years later the Dudley Brewery Co. leased land along the Birmingham Road at Burnt Tree from the Earl of Dudley on a 120-year lease from December 1805 at a yearly rent of £5. The company directors were James Bourne, a solicitor, Joseph Royle, a maltster, Thomas Wainwright, a surgeon and Thomas Hawkins, who had financial interests in the glass industry and the meat trade. The brewery was constructed soon afterwards. Three storeys in height, of conventional design for the period, with two funnelled malt houses to the rear, it was a ten-quarter in size, capable of producing up to 300 eighteen-gallon barrels

A late seventeenth-century inn that would have brewed its own beer.

The Dudley Brewery, established in 1805.

each week. The brewery is first listed in *Holden's Triennial Directory* of 1809-11 as the Dudley Porter & Ale Brewery Co. The brewery traded successfully and without competition until 1820. In that year the Dudley New Brewery at Kate's Hill was established by Henry Cox.

There are no constructional details of the new brewery, or its capacity, but in 1830 an additional building was constructed at the site, with a ten-quarter brewing capacity. Whether this was additional, or in place of the original building, is unknown. Meanwhile in West Bromwich a third common brewer had begun production. Abraham Fisher is the first of West Bromwich's brewers to be listed. He appears in *Pigot & Co.'s Staffordshire Directory* of 1828. Without a trade entry for brewers, he is listed under Maltsters as a 'maltster and common brewer' of Greets Green. Upon his death, his widow Sarah continued running Greets Green Brewery, while brother-in-law Jesse Fisher had established the Stoney Lane Brewery, off Sandwell Road, sometime before 1838. This brewery was later taken over by Heelas & Co., who renamed it the Burton Brewery. They continued trading up to 1865.

In 1827 the company running Old Dudley Brewery was restructured as James Bourne & Co. and soon after as Bourne, Cannon & Co. Thomas Dawes joined the firm in 1834-35, and he appears to have been appointed as manager. In an effort to expand the business, sales offices were established at Birmingham under Richard Telford and at Wolverhampton under John White. The company evidently began experiencing financial problems as former partners withdrew (or were ousted) from the business. This led to underfunding and an inevitable takeover. At the end of 1836 Joshua Scholefield jr was put in to run the Old Brewery.

Dudley New Brewery was also experiencing problems In September 1840 Henry Cox & Co. announced the break-up of the company. On 3 June 1841, Dudley New Brewery was put up for sale by auction. The reason why these early breweries appear to have collapsed so suddenly appears to be directly or indirectly related to the unfair competition engendered by the Duke of Wellington's Beerhouse Act of 1830. In 1830 there were 107 public houses, hotels and inns in Dudley. Within five years of the Act, a further 120 retail breweries had come into being within the borough.

TWO

RETAIL BREWERS AND HOME-BREW HOUSES

On 10 October 1830, the Beerhouse Act introduced by the Duke of Wellington came into force. The Act was designed to combat the evils of gin drinking, which had led to excessive drunkenness, depravity, extreme poverty and a weakening in the national character. The new Act encouraged people to take up the drinking of wholesome English beer. It permitted people, after purchasing a £2 license, to brew and sell beer from their own premises. Within the year 25,000 licenses had been issued nationwide and the worst excesses of drinking Holland gin were swept away. The government also managed to pocket £50,000 into the bargain, a not inconsiderable sum in those days.

Brewing in the home, for home consumption, had always been practised in the Black Country, even before the Industrial Revolution. For many the introduction of Wellington's Act meant a release from the grind of the coal mine or the foundry and the nail and chain shop. Here was an opportunity to set up in business for oneself, and for a very moderate investment. It had its pitfalls for the unwary though, and of course there was fierce competition. Bridgen's *Directory of Wolverhampton and District* for 1833 indicates the proliferation of small retail brewers responding to Wellington's Act. From there being none in 1828, expansion was swift:

Bilston	38
Sedgley★	53
Wednesfield	3
Willenhall	7
Wolverhampton	75

★ includes Coseley, Gornal Wood & Upper Gornal

With such numbers it made sense for new brewers to be cautious; to not give up the day job, so to speak. John North of Bilston, a retail brewer in 1833, was also a glazier and painter; William Wheeler was a japanner; William Pyatt was a carpenter; Richard Pedley of Willenhall was also a shoemaker, as was Edward Glover of Gospel End. Edward Guest was a gardener and James Hughes was a butcher. Though it was certainly easier than working down a mine, brewing was still very labour-intensive and time-consuming. It took between two to three hours to start up

Mrs Pardoe outside the Old Swan, 1972.   *Old Swan beer mat.*

the brewing process. All vessels had to be scrupulously clean before brewing could take place. The copper in the brew-house was lit and filled with twelve gallons of water (this being the usual half-brew), and the water brought to the boil. The boiling liquor was then transferred to mashing tubs and allowed to cool to 80 degrees. Meanwhile the copper was filled with another twelve gallons and the process repeated to eventually fill two barrels with twenty-four gallons of beer. Two or three pecks of malt were added to the mash tub and stirred well. The mash tub was then covered with clean sacking or some other similar cloth in order to draw, or extract, the malt. This was left to stand for three hours or so. The second half of the brew was similarly prepared. After three hours the liquor was transferred back to the copper and four to six ounces of hops were added. The brew was brought to the boil and allowed to simmer for two hours, being constantly stirred. Two to three ounces of black malt was then added if a darker brew was required. The beer was then strained into a cooling vessel and allowed to cool to 80 degrees then returned to the mashing tub. Half a pint of brewer's 'balm' (live fermenting beer left over from a previous brew) was added. The brew was then covered and left for a week until the 'balm' had done its work, and had dropped to the bottom of the vessel. Then the beer was put into barrels and two three-quarter pounds of sugar and half a pint of finings were added to each twelve-gallon barrel. The beer was usually clear in four to seven days and was ready for drinking. Up until the outbreak of the First World War the original gravity (OG) was 1060, which made it the second strongest brew in the country. Traditionally in the Black Country the beer was a light brown coloured mild, being darker than bitter but lighter than the commercial mild of the big brewers.

While some contented themselves with brewing and selling beer locally, the more enterprising license-holders turned their houses into public houses, which were known as home-brew houses. Many survived into the mid-twentieth century before being taken over by the bigger breweries. Perhaps the most well known of the survivors is the Old Swan in Netherton. The Old Swan, affectionately known as Ma Pardoe's, was during the 1970s one of only four home-brew house remaining in England. Among serious beer drinkers it had an almost cult status – and still does to a certain degree. The Old Swan dates back to 1835, and possibly a little earlier. Thomas James is its earliest recorded landlord. There is every reason to believe that he brewed on the premises. The present pub and brewery date from 1863, when the terrace of houses it belongs to were built. The earliest surviving deed relating to the Old Swan, dated 22 April 1863, records that the premises included a brew house, stables, outbuildings and gardens. Its then owner was John Young. In 1872 Thomas Hartshorne, a Gloucestershire man, purchased the pub and its brewery. He and his family held it until 1964. In 1932, Tommy Hartshorne, Thomas' son, offered the tenancy to Frederick and Doris Pardoe, who were then running the British Oak in Sweet Turf, Netherton. This was another home-brew house built in 1861. The couple were hoping to get the tenancy of the Ward Arms, along the Birmingham Road in Dudley. It was a rather prestigious pub, with tennis courts and a bowling green, opened in April 1927. As a temporary measure the Pardoes agreed to take on the Old Swan and moved into it in 1931. The original bar as they knew it remains virtually unchanged. It has a splendid enamelled ceiling, its centre depicting a swan. To the left of the door is an old stove and chimney, and away to the right is an old weighing machine.

When they moved in to the Old Swan, the Pardoes inherited its brewer, Ben Cole. He was later succeeded by Solomon Cooksey, and he in turn by his son, George. In an interview with *Birmingham Mail* reporter Leon Hickman (2 February 1977), Doris revealed that 'George took over from his father, not that he had much choice in the matter. His dad refused to teach any of my father's relations how it was done. He passed his secrets on to George Cooksey and no one else. Now George's son is learning the trade.' The beer that the Cookseys brewed was a cross between mild and bitter (though the *Good Beer Guide* always referred to it as a bitter). It had the light golden colour of bitter, but the malty taste of mild. It was brewed at an OG of 1032 – quite low in alcoholic content, but an ideal session beer. The Old Swan's brewery had a capacity of twenty-eight barrels a week. Its fermenting vessels were made of unlined timber.

When Fred Pardoe died in 1952 at the age of fifty-nine, his widow Doris (a teetotaller would you believe?) took over the running of the Old Swan. In 1964 she bought the freehold of the property, thus ensuring its status as an independent brewery. By the late 1970s, thanks to CAMRA, Doris Pardoe had become a legend in beer-drinking circles. It was not unusual when popping in for a swift half to discover her signing beer mats for devotees who might have come some twenty miles or more for a drink at her famous house. Doris Pardoe died on 1 April 1984. Her years matched the century. Daughter Brenda and son-in-law Sid Allport took over, but within a year the Old Swan was put up for sale. There then followed a troubled few years with the pub changing hands a couple of times. Brewing also ceased for a short period, and the original 'balm' was lost. The Old Swan was extended to side and rear in a bid to make it more profitable. Luckily it has overcome its difficulties and brewing has resumed. Its 'Original' now has an OG of 1034.

Sarah Hughes, who also lived to be eighty-four, was another Grand Dame of Black Country brewing. Then a widow, she became co-owner of the Beacon Hotel in Sedgley in 1921. She was then fifty-four years old and at an age when most people would have been looking forward to retiring. The Beacon had been built for Abraham Carter in about 1865. Upon his death his widow Nancy took over the running of the hotel until her death in 1890. Joseph Richards succeeded her as licensee, followed by John Baker, and in 1921 by James Fellowes and Sarah Hughes. Fellowes, who is listed as a brewer from 1914-21, seems to have been John Baker's brewer, but appears to have relinquished all control of the day-to-day running of the business after 1921.

In the small tower brewery behind the Beacon, Sarah began brewing a strong-flavoured ale called Dark Ruby. It was in the tradition of the old-style Black Country milds. Upon her death in 1951 her son Alfred took over, but in 1958 brewing stopped at the Beacon and beer was bought in. In 1987, Sarah's grandson, John Hughes, re-established the brewery after, it is said, the original Dark Ruby recipe, which it was thought was gone forever, was rediscovered in an

SARAH HUGHES BREWERY

BILSTON STREET, SEDGLEY

ORIGINAL HOME BREWED ALES ON-SITE SINCE 1921

Drink **DARK RUBY MILD** (OG 1058) & **SEDGLEY SURPRISE** (OG 1048)
at our attached Brewery Tap **THE BEACON HOTEL**
- situated 500 yards from Sedgley Bull Ring -

Watch beer being brewed in an original Black Country Victorian Tower Brewery

TELEPHONE: SEDGLEY (01902) 883380

Sarah Hughes' Brewery.   *Advertisement for Sarah Hughes' Brewery.*

old cigar tin. He revived the Dark Ruby Mild (OG 1058) and also began brewing a draught pale ale called Sedgley Surprise (OG 1048) under brewer Peter Hickman. The original brewery plant was restored and renovated, and an additional floor was added to the existing brew-house. On the gable end of the building is painted 'BREWERY'. Above the doorway entrance to the brew-house is a small sign which reads, 'Sarah Hughes, 1921, Home Brewed Ales'.

Like the Old Swan, the Britannia Inn at Upper Gornal developed from a terraced house when a former nailer, John Dukes, purchased a license under Wellington's scheme and began brewing. Dukes converted his house into a home-brew house and upon his death his widow Hannah and son Richard continued the business. The Britannia was acquired by Joseph Round Cartwright, who sold it on to butcher Henry Perry in 1864. The family were associated with the Britannia for the next 127 years. Henry's son, Louis Peacock Perry, and wife Sarah, took over the pub and brewery following Henry's death. Upon Louis' death his widow and married daughter, Sally Williams, ran the pub until 1942 when Sarah retired, leaving daughter Sallie and her husband Wally to run it. Older drinkers in their eighties recall that Perry's brewed a traditional Black Country dark sweet mild, which pre-war would have been OG 1060. Brewing ceased at the Britannia in 1959. Like so many Black Country pubs, the Britannia had an alternative name. Through the 1970s and '80s the pub was also known as 'Old Sal's' after its popular landlady, Sallie Williams, who died in 1991. The license was briefly taken up by John Burrows, but in November 1992 Phil Bellfield took up the license and brewing was resumed at the Britannia in 1995, using a three-barrel plant. Batham's took over the Britannia Inn in 1997 and brewing ceased once more.

It was Batham's who were instrumental in introducing another legendary Black Country brewer. Thomas Booth, or Tommy as he was more familiarly known, was a near contemporary of Fred Pardoe. An Old Hill man, he had moved to Netherton where he had found work at the New Golden Colliery. In 1916 he approached Daniel Batham jr regarding the vacant tenancy of the King William Inn in Cole Street. Batham agreed, and Booth took up the tenancy. Booth was an exponent of the 'long pull', a system whereby the landlord would top-up a partially drunk pint, thus ensuring customer loyalty. This loyalty, apart from making him a champion of the working man, also enabled him to buy the nearby Blue Pig in St Andrew's Street five years later. He paid £2,300 for the premises, which included a disused brewery at the rear. Booth rebuilt the brewery and, with the assistance of Solomon Cooksey sr, began brewing at the Blue Pig. Cooksey produced a light mild, which, given the secrecy of the family, must have been very

Tommy Booth's Red Lion home-brew house, Gornal Wood.

similar, if not the same, as that later brewed at the Old Swan. In addition Cooksey brewed a Winter Ale and a strong Christmas Ale.

Booth now expanded the business by buying the Sampson & Lamb in Halesowen Road, Netherton, and an off-licence in Bald Lane, Lye. In August 1935 Booth bought the Red Lion Inn and brewery in Abbey Road, Gornal Wood, from William Elwell for £3,650. The pub had a good reputation. It had been run by a local character, Thomas Malpass, for over thirty-eight years. Booth rebuilt and extended the old brewery and moved all production to here. Booth now began to brew a much darker, malty tasting mild, which he supplied to his stable of nine tied houses and off-licences, plus free-trade outlets. His beer was also bottled by Bird's Crown Brewery at Bloxwich and later by Holden's of Woodsetton.

After the death of his first wife, Booth married widow Annie Round. In 1939 the couple moved to Pensnett, where Booth opened a new brewery in Corbyn's Hall Lane. Booth's daughter Charlotte, by his first wife, took over the running of the Red Lion. In November 1942, due to war-time shortages, Booth sold the Red Lion to Julia Hanson & Sons of Dudley. Tommy Booth died in October 1952. The Corbyn Hall Brewery was sold and subsequently demolished. Booth's Kinver pub, The Plough & Harrow, was sold to Batham's.

Batham's, which had given Tommy Booth his first experiences as a licensee in the Black Country, was founded in 1877. Now in its fifth generation of family ownership under Tim and Matthew Batham, the company celebrated its 125th anniversary in 2002. The firm was established by Daniel Batham at Netherton, though rumour has it that Daniel's wife took up brewing in the home some ten years prior. In 1882 Daniel bought his second pub, The White Horse in Cradley High Street, and in 1904 the King William in Cole Street, Darby End, Netherton. In 1905 Batham's bought The Vine in Delph Lane, Brierley Hill. The pub – which had formerly belonged to Caroline Cox – was to become their flagship. There had been some hesitation on their part in buying the pub – the land was vulnerable to mining subsidence. In

Above: *A Batham's pub display board.*

Left: *An advertisement for Batham's.*

fact the original Delph Brewery, a little further down the road, had almost entirely fallen down. The cause was the shallow Dock-o'-Nine pit, which ran beneath it. Nevertheless the family went ahead and bought the house. In 1911 The Vine was totally rebuilt and renamed The Vine Hotel. A model tower brewery and plant was built behind it. Like many Black Country pubs The Vine also has another name – 'The Bull & Bladder'. Painted above the frontage, Batham had inscribed, 'Blessings of the Art, Thou Brew'st Good Ale'. This was a quote from Shakespeare's *Two Gentlemen of Verona*. The motto was subsequently changed to 'Blessing of Your Heart: You Brew Good Ale'. Brewing was moved from the other pubs to the new Delph Brewery. There then featured a period of consolidation, the firm having overstretched itself. In 1922 the White Horse was sold to Hezekiah Dunn for £3,455. This eased financial constraints. The following year the company bought the Royal Oak in Lye, Stourbridge. In 1926 they bought three more pubs, the Bird in Hand, Hagley Road, Old Swinford; the Brickmaker's Arms, Dudley Road, Lye; and the Spread Eagle in Brierley Hill. The first two were later sold to the Wolverhampton & Dudley Brewery (W&DB) in 1940. Of all their purchases the most profound was their lease of the Swan Inn, Chaddesley Corbett, in 1951. Up to this point Batham's had been brewing a mild ale at OG 1036. (ABV 4.5%). It was a pleasant enough pint, a sweet brown ale with a hoppy, almost fruity taste. It did not suit the tastes of the drinkers in this part of Worcestershire though. They were used to bitter. As a consequence Batham's were obliged to brew a bitter for them, or risk losing their custom. They came up with a straw-coloured best bitter, with an OG of 1043 (ABV 4.5%). It was an immediate success and was introduced to all of Batham's houses. Later the company began brewing a Christmas ale called XXX, with an OG of 1064 (ABV 6.5%) harking back in strength to the Black Country beers of old. Following earlier setbacks, the company has once more expanded and has nine tied houses. Its latest is the Y Giler Arms, Betws-y-Coed, North Wales.

Just down the hill from Batham's was the brewery of J.P. Simpkiss in Brettell Lane. The family's connection with home-brewing began in 1854 when former potter William Simpkiss turned licensed victualler, following his purchase of the Potter's Arms in Rock Street, Brierley Hill. In

1869 he retired, and his brother Robert took over. That same year William's twenty-five-year-old son, William Henry, having borrowed £700 from his father, bought the Royal Oak at Round Oak and a piece of land behind it. A few years later he built a brewery on the site. William sr died in 1871 and the Potter's Arms was sold soon after, presumably to pay off any debts that had accrued. By 1890, with production at 250-350 barrels a week and already supplying the free-trade market, William Henry bought up three licensed houses: The Red Cow, Brierley Hill; The Fountain Inn, Bradley Street, Pensnett; and The White Hart Inn, Brewery Street, Wordsley. Simpkiss' expansion attracted predators. North Worcestershire Breweries, formed in 1886, approached Simpkiss in 1896 with an offer of £20,000 for his brewery and eleven freehold tied houses. He accepted, and retired to become a gentleman, dying in 1905. As is so often the case, North Worcestershire Breweries were only interested in the pubs. The Royal Oak Brewery was sold on the following year to Elwell Williams, who renamed it the Town Brewery.

In 1903, twenty-nine-year-old Joseph Paskin Simpkiss, William Henry's son, bought the Swan Brewery in Evers Street, Quarry Bank. It had formerly belonged to John S.A. Bell, a Londoner, who had paid £1,900 for it in 1894. Simpkiss registered his new company as Home Brewery (Quarry Bank) Ltd. The company were quoted as having capital of £10,000. The money had presumably come from father William Henry. Joseph began supplying the free trade, but as money became available the company began buying up houses as they became available, until they had a stable of twenty-three tied houses. By the outbreak of the First World War they were brewing 300 barrels a week. The war had its impact on the brewery, as presumably with all similar small breweries. Brewing was restricted by the Government and duty on beer rose from seven shillings to twenty-three shillings a barrel. Coupled to this was the cut in opening hours. Profits dropped. In 1916 a bizarre court case ensued. For whatever reason, Simpkiss had inadvertently signed over control of the business to his office manager, William Thomas Clewes. Clewes took Simpkiss to court, and won, effectively removing the former Managing Director

Simpkiss' Brewery, Brierley Hill.

Commemorative beer label.

from the company. Clewes became Managing Director and took Head Brewer William Proctor on as a partner.

J.P. Simpkiss, having been deposed, found work as a travelling salesman for Smith & Williams' Town Brewery, Round Oak. He saved and borrowed money until at last he had enough to buy another home-brew house. On 26 April 1919, Simpkiss bought the Foley Arms for £3,000. This pub dated from pre-1851, when John Davies had been its licensee and it had borne the name of the Wellington Arms. The brewery was in what was later to become the lounge of the Foley. By 1924 Simpkiss was producing 145 barrels a week. Expansion now became the order of the day. He began leasing, then later buying, pubs as the money became available. Improvements were also made to the old brewery to bring it up to date.

In 1926, son Dennis Simpkiss joined the firm. Perhaps in his honour, a new brewery was built at the rear of the pub and named the Dennis Brewery in 1934. It was a five-quarter brewery, capable of producing 250 barrels a week. Brewing in the Foley Arms was discontinued and the room became the lounge of the brewery tap. The company began producing a well-hopped malty bitter at OG 1037 and a fruity and full-bodied old ale at OG 1050. A bottling plant was also introduced. In 1955 the company amalgamated with Johnson & Phipps Ltd of Lichfield Street, Wolverhampton. It was in their mutual interests. Johnson & Phipps Brewery was under compulsory purchase by the Council, who wished to demolish it for town improvements. The name of the new company became J.P.S. Brewery, taking the three initials from the two companies. Brewing was transferred to Brierley Hill, but the Wolverhampton company retained control over its seventeen tied houses. When Director Alan Phipps retired in 1969, his pubs were put up for sale. Simpkiss bought four of them. The remainder were sold to W&DB. In 1977 the company reverted to its original name. Dennis Simpkiss died in 1981 and son Jonathan became Managing Director. Barely had he settled into the job when in 1984 Greenhill-Whitley put in a take-over bid. It was rejected, Jonathan declaring, 'Local people do like to drink local beers – and that's why we're staying in business', the *Express & Star* reported. In July of the following year the offer was increased and the company agreed to the sale of the brewery and its fifteen tied houses. The brewery, not being needed, was shut down with undue haste and shortly after demolished. Twenty jobs were lost.

Holden's of Woodsetton was founded by a former shoemaker, Edwin Alfred Holden. At the age of twenty-two he married Lucy Blanche Round, daughter of Ben Round, landlord of the Trust in Providence, Netherton. He talked Holden into taking on a pub, and persuaded Atkinson's of Aston, Birmingham, to appoint the Holdens to the tenancy of the Britannia Inn, Netherton, in 1898. Over the course of the next twenty-two years the couple moved some four times, gaining considerable experience. In 1915, while running the Summer House in Woodsetton, they bought the Park Inn in George Street, Woodsetton. In April 1920 they gave up the Summer House and moved into the Park Inn. It was a home-brew house and with it came a brewer, Harry 'Ossie' Round, a relative of Lucy's. Behind the pub was Atkinson's malting house, which in the fullness of time was to become the Holden's Hopden Brewery.

Edwin Alfred Holden died in August 1920 and his wife took over the license. She held it until 1938 before handing it over to her son, Edwin Alfred jr, more popularly known as Teddy. During her stewardship she began to expand brewery production and bought up a second house, the Painters Arms. Ossie Round retired in 1934 and was succeeded by Harry Field, another family member. Son Teddy, meanwhile, was at Birmingham University studying brewing. Upon graduation Teddy was given the Painters Arms to run, to gain experience. In June 1938 Teddy and his wife Clara moved into the Park Inn. Lucy had died in the previous month, aged just sixty. In 1939 Atkinson's old malt house was taken over and a new brewing plant was installed.

Holden's Brewery, Woodsetton, 2004.   *Holden's beer mat.*

In 1944 Holden's bought their third house, the New Inn, Coseley. The following year a bottling plant was installed at Woodsetton and the firm began bottling not only their own brew, but also the home-brew of other pubs. Holden's continued buying pubs when they could afford them. It was steady and sensible growth. By 1960 they had six tied houses. In 1961-2 the old brewery was rebuilt and enlarged. The firm was registered as Holden's Brewery Ltd, Hopden Brewery, Woodsetton, in 1964, with a capital of £75,000. Teddy's son, another Edwin, who had studied at McMullen & Sons Hertford Brewery, joined the firm in 1965. Another pub, the Cottage Spring in Wednesbury, was bought in 1966. By the 1970s Holden's was producing more than 240 barrels a week and had a stable of ten tied houses. Edwin became Managing Director upon the death of his father in January 1981. Slow but sure, the company added to their tied houses. By 1983 Holden's had sixteen tied houses, making it the largest of the Black Country home-brewers. Edwin Holden died in December 2002. The brewery's latest acquisition had been the former Codsall Station, which was converted into a pub, bringing their estate up to twenty-two tied houses. In addition they had acquired a further sixty-one outlets. Currently (as of 2005), Holden's brew five regular beers, Black Country Bitter (3.9%), Black Country Mild (3.7%), Golden Glow (4.4%), Special Bitter (5.0%) and XB, also known as Lucy B. (4.2%). In May of 2005 they introduced a new brew, Will 'o' Wisp, an amber premium beer (4.7%).

There was one Black Country brewery that made the transition from home brewery to common brewery. John Rolinson and his extended family were part of a brewing tradition in Netherton. John himself had formerly been a bricklayer, who in 1852 bought the Bricklayer's Arms in Church Street. The family's story was to be a roller-coaster ride of incredible success destroyed by greed and irresponsibility. By 1871 Rolinson had moved to the Five Ways Inn at 45 St Andrew's Street. The house took its name from a road sign with five pointing arms. On the land behind, Rolinson built a brewery. A description of it appears in the March 1896 edition of the *Brierley Hill Advertiser*. It is described as a '15 Quarter Brewery, known as the Five Ways Brewery… adjoining the corner double-fronted fully licensed public house, the Five Ways Inn…' In addition the brewery had the fifteen-quarter Malt Mill House in Northfield Road, near the Loving Lamb. The brewery was capable of producing 850 barrels a week.

The company put incredible strain upon its finances by buying up five pubs in Netherton in quick succession from which to sell its beer. Properly financed and thought out, this was the way to progress. Unfortunately the purchases left them financially stressed. In order to raise more money they started to defraud the Inland Revenue. In accordance with section 20 of the

Above left: *Advertisement for Holden's, c. 1955.*

Above right: *Advertisement for Holden's, c. 1985.*

Right: *Rolinson's entry in the Brewery Manual, 1899.*

Below: *The old Five Ways Brewery and tap.*

854

ROLINSON (JOHN) & SON, LIMITED.

Directors—

DANIEL ROLINSON (Chairman). | EDWARD WEBB, J.P. | JOHN BARTHOLOMEW GOMM.

Secretary—REGINALD W. BOOTH.
Auditor—T. H. GOUGH.

Office—THE FIVE WAYS BREWERY, NETHERTON, DUDLEY.

Authorised Share Capital, £100,000, divided into 3,000 6% Cumulative Preference Shares of £10 each, and 7,000 Ordinary Shares of £10 each. 4½% First Mortgage Debentures, £60,000 (redeemable, on June 1st, 1916, at 110%). Interest payable June 1st and December 1st. Preference Dividend payable same dates.

VALUER'S REPORT.

The following valuation by Mr. A. W. Dando, Auctioneer and Valuer, of Dudley, appeared in the prospectus:—
"18 Wolverhampton Street, Dudley, February, 1899.
"To the Directors of John Rolinson and Son, Limited (Messrs. D. Rolinson, E. Webb, and J. B. Gomm), The Five Ways Brewery, Netherton.
"DEAR SIRS,—Agreeably with your instructions, I have personally inspected and surveyed the Five Ways Freehold Brewery at Netherton (the greater part of which has been recently rebuilt and refitted with a modern 20-quarter plant), together with 47 freehold and nine leasehold and annual tenancy licensed houses (of which 32 are fully licensed, 20 beer licensed, and four off-licensed), together with two freehold maltings, stabling, offices, 54 shops and dwelling-houses, and several plots of building land, more particularly described in the Schedule hereto annexed. The houses have been carefully bought, are for the most part within easy carting distance, many being quite close to the Brewery, and nearly all are in the hands of genuine tenants at very moderate rentals.
"In my opinion the foregoing freehold and other properties, as at present licensed, are of the value of Ninety-six thousand two hundred pounds.
—Yours faithfully,
"ALFRED W. DANDO."

ACCOUNTANT'S CERTIFICATE.

Mr. T. H. GOUGH, Chartered Accountant, of Dudley, certified that the average profits of the business for the eighteen months ended December 31st, 1898, amounted to £8,536 16s. 3d. per annum.

Remarks.—The Company was formed on March 8th, 1899, to purchase the Brewery business of a Company with a similar title, trading as The Five Ways Brewery, Netherton, Dudley. Over 50 Licensed Houses were included in the Purchase. The Purchase Price was £110,000, payable as to £60,000 in cash, £25,000 in Ordinary Shares, £5,000 in Preference Shares, and £20,000 in Debentures. The Public issue consisted of £40,000 in 4½% Mortgage Debentures, and 1,000 6% Preference Shares of £10 each, both at par.

Brewing Act, Rolinson had been provided with a book in which to enter the date and hour of each brewing, and to enter all the materials used by him. On the afternoon of 3 December 1884 two Excise men were present at a brewing in which 672lbs of sugar had been dissolved and added to the brew in accordance with the entry. Apparently acting upon information that fraud had taken place previously, one of the Excise men hid himself in the brewery to watch what might happen. A brewery employee, a man named 'Bill', added another bag of glucose to the brew. The produce of that day's subsequent brew was discovered to be 222 gallons in excess of what they had declared, if the brewer, John Pullin, had adhered to the quantity of materials used. A previous count of the sugar was checked and discovered to be one bag short. John Rolinson's son, Daniel, claimed that he had sold the missing bag to a Netherton man named John Johnson for fifteen shillings. Evidence came out in court that Johnson was financially linked to the brewery, and his testimony was suspect, though nothing could be proven. There was no getting away from the fact that the entry in the book was false. John Rolinson was fined. Son Daniel, who had worked for his father from the age of fourteen, was appointed a full partner in the firm later that year and effectively took over the day-to-day running of the business. John Rolinson died in 1896. He left the company close to bankruptcy. Daniel had no choice but to put the entire business up for sale by auction. With a breathing space, he succeeded in raising the first of several mortgages and was able to buy back the brewery and its six tied houses.

Another member of the family, Richard Rolinson, kept the Old Pack Horse in Hill Street, Netherton. It was a home-brew house with a brewery in the yard behind. He over-reached himself in his attempted expansion, and was declared bankrupt in August 1885. The *Brewers' Journal* (15 September 1885) listed his debts at £1,209. His biggest creditor was G. Thompson & Son of Dudley, to whom he owed £391 3s 6d. He owed £60 to a relative, John Rolinson. Richard Rolinson suffered a further mishap on 26 March 1895 when his Netherton New Brewery was almost destroyed by fire. The damage was estimated at several thousand pounds, 'but happily', as John Richards charitably recounts in his *History of Holden's Black Country Brewery*, 'the property was insured'. You bet it was. Rumours were rife at the time, but nothing could be proved.

Meanwhile the Five Ways Brewery was lurching towards another crisis. Daniel was obliged to sell off two of his tied houses in order to raise £2,000. As the brewery struggled to become financially viable, Daniel's lifestyle – in line with his election to public office as a councillor – plunged it once more into crisis as he went on a spending spree to maintain the illusion that he was a man of means. In order to raise more money, the company went public, and in March 1899 was registered as John Rolinson & Son Ltd with a capital of £100,000. Advertisements were placed in the newspapers and the company diversified into coalmining, brick making and land. Daniel bought hotels and a farm. He became a country gentleman. But all was not well. Over the years the brewery had been heavily mortgaged. Daniel was himself declared bankrupt in 1910. What was required was a steady hand at the tiller. An Extraordinary General Meeting was called in 1912 with the purpose of injecting capital into the business by allowing Wolverhampton & Dudley Breweries to buy extra shares. It was stressed to the shareholders that this was not a takeover, but the use of outside financial expertise, to steer the company out of its temporary crisis. Despite the semantics it was a takeover, but the company was allowed to continue trading under its own name until December 1925, when it went into voluntary liquidation. Its tied estate of fifty-eight houses and off-licences, already selling W&DB's beer, were taken over completely.

This was the era of the big brewer, but somehow it is nice to know that some home-brew pubs have survived. They are small in scale, and somehow more in harmony with their customers. Would you expect to find the Managing Director of W&DB in the brewery tap, signing beer mats?

The list of local brewers below was compiled from Licensing Justices lists, Kelly's Directories, The *Brewers' Journal* and newspaper articles. Entries are given under town and village. Home-brew houses are also listed where the house brewed only for its own clientele. Where these houses were known to have expanded to supply other houses, either their own tied houses, or the free trade, these breweries are listed as common brewers in the main list.

BILSTON

Joseph Adams, Temple Street, 1833. A retail brewer.

Joseph Adams, Skidmore Road, Bradley, 1930.

George Adderley, Union Street. A retail brewer, first listed in Bridgen's Directory of Wolverhampton for 1833.

Thomas Banks, Catchem's Corner, 1833.

John Bate, Oxford Road, Newtown, 1828. A retail brewer.

Bilston Brewery, 128 Oxford Street. Joseph Marsh, brewer, 1900-06.

Oliver Brown, 92 High Street, 1921-30. Mrs S.J. Brown, 1935.

J. Bullock & Cooper, Wynn's Court, High Street, 1833.

William Butler, grocer, brewer, ale and porter and hop merchant, New Village Brewery (Priestfield). See: Butler's Springfield Brewery in main listings.

James Clapperton, Oxford Street, 1833.

George Cox, Broad Street, 1900-10.

William Darbey, 15 Coseley Street, 1914-21.

William Davies, Temple Street, 1833.

Thomas Dimmack, Oxford Street, 1833.

Samuel Downing, Church Street, 1833. Also a millwright.

Druid's Head, Caddick Street, Hurst Hill. A home-brew house run by Joe Adams, whose family held the license for over 100 years. Better known as Flavells in its latter years, having been run by Joe Flavell for fifty years. It closed in November 1971.

William Edwards, Temple Street, 1833.

D. Evans, Market Street, 1864. A retail brewer.

Mrs Susan Evans, 92 High Street, 1914.

John Fanton, Market Street, 1833.

Isaac Fellows, 25 Broad Lane, 1923.

Joseph Fellowes, High Street, 1833.

Thomas Fieldhouse, Ettingshall Lane, 1833.

Thomas Fieldhouse, Wright Street, Bradley, 1914.

Joseph Fletcher, High Street, 1828.

William Fletcher, Shropshire Row, Bradley, 1833.

Arthur George, Salop Street, Bradley, 1914.

Henry George, 19 Cross Street, Bradley, 1914-21.

Joseph Griffiths, Salop Street, Bradley, 1914-21.

Edward Hand, Canalside, Bradley, 1833.

George Harbach, 55 Bridge Street, 1921-26.

Horace Henry Harbach, 20 Market Street, 1926.

Walter William Harbach, 12 Walsall Street, 1914-23.

John Harrison, Coseley Street, 1833.

Thomas Hartshorn, Ettingshall Lane, 1833.

Haskew & Co., 30 Gozzard Street, 1908.

Thomas Hatton, Coseley Street, 1833.

William John Hawkes, 7 Cross Street, Bradley, 1914-39.

William Homes, Moxley, 1833.

Mrs Sarah Hughes, 179 Oxford Street, 1914.

Thomas James, Lichfield Street, 1833. Also a moulder.

William James, Shropshire Row, Bradley, 1833.

Mrs Betsy Jeavons, Bank Street, Bradley, 1914.

Thomas Jones, Union Street, 1833.

John Leadbeater, Coseley Street, 1833.

Josiah Longmore, Shropshire Row, Bradley, 1833.

Joseph Lowe, Temple Street, 1828.

John Maddocks, Stow Lane, 1833.

Samuel Mann, Ettingshall Lane, 1833.

Joseph Wilkes Marsh, Bilston Brewery, 120 Oxford Street, 1878-1904.

John Maybury, Catchem's Corner, 1833.

George Meese, 11 Temple Street, 1914-21.

Isaac Millard, Deepfields, 1921-23.

Thomas Morris, Oxford Road, Newtown, 1828. Retail brewer.

John Thomas Mottram, 20 Market Street, 1921-23.

John North, Oxford Street, 1833.

John North, Union Street, 1833. Also a glazier and painter.

Simon Paul, Hill Street, Bradley, 1914-23. Daniel Paul, 1930.

Peck & Kerrison, Wesley Street, Bradley, 1888-1916.

Thomas Perkins, 176 Oxford Street, 1864. Retail brewer. Mrs S. Perkins, 1868.

George Price, Mill Street, 1888-92. Mrs Elizabeth Price, 1900.

William Pyatt, Church Street, 1833. Also a carpenter.

Royal Exchange, 58 High Street. Established as a home-brew pub by butcher William Fellowes in 1861. Bought up by J.& J. Yardley of Bloxwich in 1899. Now owned by Holden's.

Joseph Shale, Cold Lanes, 1833.

William Shale, Wolverhampton Street, 1833.

William Simms, Hall Street, 1833.

Spread Eagle, Lichfield Street. Formerly two cottages knocked into one and dating from the early nineteenth century. It is once more a home-brew pub producing Skittain Ales. See also: British Oak, Salop Street, Dudley.

Thomas Stringer, Oxford Road, Newtown, 1828.

Edward Ward, Skidmore Row, Daisy Bank, Bradley, 1914. Thomas Ward, 1921. Edward Ward once more in 1923.

William Wheeler, Crown Street, 1833. Also a japanner of metal.

Benjamin Whele, Stafford Street, 1833. Maltster and retail brewer.

Joseph Winsper, Temple Street, 1833.

BLACKHEATH

Thomas Chapman, Halesowen Street, 1884.
Alfred Cox sr, Long Lane, 1878. Alfred Cox & George Major, 1880. Also Alfred Cox, Ashley Hotel, Long Lane, 1884.
William Darby, 1884. Mrs Emily Darby, 1888.
Jeremiah Downing, The Nimmings, Cakemore, 1904.
Frank Garrard, 1892. Garrard Brothers, 1892, Frank Rochfort Garrard, Long Lane, 1896.
Andrew Harris, 1884-88.
Mrs Richard Merris, Mincing Lane, 1914.
Mrs Emily Parkes, Olive Street, 1921.
Edward Sturman, 392 Long Lane, 1884-1935.
William H. Taylor, High Street, 1914-21.
Isaac Troman, 392 Long Lane, 1939-40.
Silas Whitehouse, Handel Hotel, 1884.
D. Willetts, White Heath, 1884.

BLOXWICH

Mrs Margaret Marshall, 614 Bloxwich Road, 1914. Alfred Marshall, 614 Bloxwich Road, 1923.
George Merrall, Stafford Road, 1914.
William Parkes, 13 Elmore Row, 1914-23.

The Royal Exchange Brewery, Bloxwich.

John Smith, Church Street, 1914-21.
Harry Tolley, Portland Street, 1914-20.
Mrs Jane Tolley, Field Street, 1914-21.
Joseph Wall, Church Street, 1914.
Frank Wilkes, 13 Wolverhampton Road, 1914-26.

BRIERLEY HILL

Benjamin Andrews, Brockmoor, 1914.
Sarah Ann Bailey, Buckpool, 1914-20.
Thomas Banks, 139 Dudley Street, 1921.
Mrs Ellen Bishop, Brockmoor, 1921.
Henry Bolton, Brettell Lane, 1914-20.
Mrs Mary Cartwright, Brockmoor, 1914.
William Cartwright, Moor Lane, 1914-35.
Thomas Clulow, Brick Kiln Street, Harts Hill, 1914-21.
Alfred Dunn, Mount Pleasant, 1914-23.
John Elton, 58 Delph, 1904.
George Elwell, Delph Brewery, 1884-88. See: Delph Brewery in main listings.
Edward Fletcher, Lower Delph, 1910.
Emma Gill, Buckpool, 1930.
John Glover, 84 High Street, Silver End, 1914-23.
William Henry Goring, Mill Street, 1921-26.
Cornelius Gorton, 1 High Street, 1914-30.
Mrs Phoebe Hartshorne, Brockmoor, 1914-23.
Mrs Maria Higgs, Moor Lane, 1914.
Home Brewery Co. Ltd, Evans Street, 1914-21.
Harry Jeavons, Brettell Lane, 1914-35.
Thomas Jeffries, Moor Lane, 1923-30.
Mrs Alice Kinsell, 92 Bank Street, 1914-26.
George Pearson, 154-5 High Street, 1914-26.
Plough Brewery, Church Street, Situated alongside the Plough Inn. Founded by Fred
   Warren in 1890. Upon his death in 1926 his widow Agnes sold the pub to J.P. Simpkiss.
Mary Rose, retail brewer, 1921.
J. Smith, Buckpool, 1921.
Spread Eagle, 2 High Street. A home-brew house run by George Pearson from 1914 to
   1926. In that year he leased the premises to Batham's. Later bought by them, they sold it
   in 1946. It closed the following year.
Swan Inn, Mill Street. A home-brew house, licensed in 1898. Owned by Edward Harley
   in 1937, he leased it to Batham's. It closed in 1939.
Archibald Edward Vale, Moor Lane, 1914-26.
Richard Wassall, Brettell Lane, 1914. E. Wassall, 1921-24.
Frank Webb, Brettell Lane, 1904-12.
Mrs Sarah Ann Wood, Bank Street, 1914-23. Robert H. Wood, 91 Bank Street, 1930-40.
Isaac Woodcock, Brockmoor, 1914.
Walter Woodhall, Delph, 1914.

COSELEY

Miss Alice Allen, Old Meeting Street, 1921.

Harry Bailey, Old Meeting Street, Walbrook, 1921.

Mrs Lydia Bailey, Walbrook, 1914.

Harry Baker, Darkhouse Lane, 1921.

Richard Bates, Cinder Hill, 1921.

William T. Baylis, Wood Street, 1930-40.

John Bryan, Hollywell Street, Hurst Hill, 1914. Mrs Mary Ann Bryan, 1921-40.

Bush Inn, Daisy Bank. A home-brew house run by Joe Adams jr from 1928-78. Bought by Holden's following Adams' death.

Job Butler, Woodcross Inn, 1935.

Mrs Sarah Chavase, Lower Ettingshall, 1878.

Edward Chesworth, Biddings Lane, 1914-21.

Richard Clee, Old Meeting Road, 1914.

William Clift, Hurst Road, Hurst Hill, 1914-21.

George Davies, Darkhouse Lane, 1914.

Druid's Head Inn Brewery, a well-known home-brew house that stopped brewing in 1971 following the death of licensee Jack Flavell.

Isaac Fellows, Broad Lane, 1914-30.

Mrs Sarah Fellows, Hall Lane, Hurst Hill, 1914. Miss Jane Fellows, 1921-23.

Joseph Flavell, Caddick Street, Coppice, 1935-39.

H. Goodridge, 50 Clifton Street, Coppice, 1935-39.

Michael Hanrahan, Hall Lane, 1926.

Ernest Holcroft, 19 Ladymore Road, 1930-40.

Thomas Holmes, Sodom, 1921.

Mrs Catherine Hughes, Clifton Street, Coppice, 1914. Dan Hughes, 1921-30.

James M'Cloud, 1833. Retail brewer.

Richard Marsh, Wood Cross Inn, 1923-30.

Mrs Honor Meddings, Hurst Hill, 1914. William Meddings, 1921-35. Thomas Meddings, Clifton Street, 1939-40.

David Millard sr, Broad Street, 1914-21.

David Millard jr, Coppice, 1914-23

Isaac Millard, Deepfields, 1926.

John Naylor, Hurst Hill, 1914-21.

Old Bush Inn, Skidmore Road. A home-brew house, established by 1840. Joseph Adams was still brewing in the 1960s. The Old Bush was taken over by Holden's in 1969.

Painters Arms, Avenue Road. A former home-brew pub in existence by 1817 and run by William Taft, who bought it from Joseph Whitehouse. He was succeeded by son John Taft, a painter (hence its name), who held the license from 1831 to 1860. The premises were acquired by Holden's in 1939, and Teddy Holden brewed here alongside his cousin, Harry Field.

Mrs Richard Price, 37 Walbrook, 1914. Mrs Clara Jane Price, 37 Walbrook Street, 1921-40.

Albert Rhodes, Sodom, 1914.

Stephen Rhodes, Wood Street, 1921-23.

Mrs Ann R. Richards, Castle Street, 1914. Joseph Richards, Castle Street, 1921.

Daniel Rowley, Dark Lane, 1914–21.

John Sharkey, Sodom, 1914.

Mrs Mary Ann Swann, Hurst Road, Hurst Hill, 1914.

William Tranter jr, Hurst Hill, 1923–40.

James Turley, 1833, retail brewer.

Mrs Mary Ward, Coppice, 1930.

Albert Williams, Darkhouse Lane, 1914–21.

John Thomas Wilson, Broad Street, 1923–40.

William Wood, Hurst Road, Hurst Hill, 1921.

Woodcross Inn Brewery, a home-brew house. Richard Marsh, 1921. Job Butler was licensee in 1935.

Mary Wright, 1833, retail brewer.

CRADLEY & CRADLEY HEATH

Alfred Aston, Providence Street, 1914–26.

James Bartlett, Dudley Wood, 1935.

Harry Bellfield, Grainger's Lane, 1914–21.

Robert Botfield, Dudley Wood, 1921–23.

George Norman Bridgewater, Victoria Brewery, Dudley Wood, 1921–39. Bridgewater later moved into motorcycle speedway and greyhound racing.

Benoni Buttery, Dudley Wood, 1914–23.

William Chilton, Furlong Lane, 1914–21.

Frederic Edward Cutler, Lyde Green, 1900–12.

Thomas Darby, Grainger's Lane, 1914. Thomas Darby & Sons Ltd, 1921–23.

Benjamin Davis, High Street, 1930.

Hezekiah Dunn, High Street, 1930–35.

Mrs Hannah Hingley, High Street, 1914.

Thomas James, Halesowen Road, Old Hill, 1940.

Mrs Esther Lane, King Street, 1914–30.

William Mason, Dudley Road, 1914.

W.H. Nock, Four Ways, 1864.

Thomas North, High Street, 1921–23.

Frank Oliver, Colley Gate, 1896–1908. W. Oliver & Sons, Colley Gate, 1914–30.

George Henry Oliver, Colley Gate, 1935.

William Partridge, High Street, 1914.

Laura L. Perry, Reddall Hill Road, 1940.

Benjamin Price, Holly Bush Street, 1914–26.

Llew Province, Lomey Town, 1914–20. Richard Llew Province, Lomey Town, 1921. Grainger's Lane, 1923–40.

Railway Tavern, Grainger's Lane. A home-brew house run by Harry Belfield from 1913 to 1932. In 1932 he leased it to Caleb Batham, of the Brierley Hill family of brewers, who thereafter supplied the house. Belfield later sold the house to Wolverhampton & Dudley Breweries.

Llewellyn Robinson, High Street, 1914.

George Roper, Colley Gate, 1914. Mrs Ellen Roper, 1921–23.

George Stafford, Colley Gate, 1921-39.

Swan Inn, Provident Street, A home-brew house run before the First World War by Harold Jasper, and alternatively known as Jasper's. It was bought up by M&B and closed in the 1970s. Later acquired by Stan Owen, it is now a Holden's house.

Talbot Brewery, Talbot Hotel, Colley Lane, 1880.

John Tandy, Blue Ball Lane, 1921-35.

Samuel Taylor, Two Gates, 1914.

William Thomas, Two Gates, 1914.

Tibbetts, Frank M., King Street, 1935-40.

William Tibbetts, Grainger's Lane, 1914.

Benjamin Tromans, Furlong Lane, 1923.

White Horse Inn, High Street, A two-storey brewery behind the pub, with a large malt and hop room. It was owned and run by Daniel Batham from 1914 to 1922. The pub and brewery were bought by Hezekiah Dunn in November 1922. He successfully ran the business until 1935. Dunn was succeeded in 1935 by John Thomas Webb.

Gilbert Willetts, Blue Ball Lane, 1939.

William Woodall, Blue Ball Lane, 1914.

DARLASTON

Thomas Allen, Willenhall Road, 1914.

Joseph Bailey, St George Street, 1914. Mrs Ann Bailey, 1921-23.

James Bayley, Catherine's Cross, 1828.

Edwin Beesley, Walsall Road, 1923.

George Bryan, St John's Road, 1923-40.

Job Butler, Catherine's Cross, 1914.

John Corbett, New Street, 1828. A retail brewer, also at Pinfold Street.

John Cotton, Smith Street, 1914.

Andrew Dawes, Spring Head, 1828.

Samuel Dowen, Blakemore Lane, 1921-23.

Mrs Sarah Dowen, Foundry Street, 1930.

William Dowen, The Green, 1914-23.

Hugh Dudley, St John's Road, 1914.

Sydney James Dudley, St John's Road, 1914.

Mrs Rose Fieldhouse, Station Street, 1914.

Benjamin Ford, Walsall Road, 1930.

Eleanor Foster, Catherine's Cross, 1828.

Martin Percy Foster, Cramp Hill, 1914-21.

Henry Freeth, Bell Street, 1914-23.

Mary Garrington, The Green, 1914.

John Green, Blockall, 1872-92.

William Handley, Forge Road, 1914.

Enoch Harper, Walsall Road, 1921.

Joseph Harper, Cock Street, 1828.

James Hitchings, Willenhall Road, 1921-23.

William Horton, Blakemore Road, 1914-23.

```
┌─────────────────────────────────────────────────────────┐
│                                                         │
│            JAS. PRITCHARD & SON,                        │
│                                                         │
│       DARLASTON BREWERY, DARLASTON,                     │
│                                                         │
│               WHOLESALE AND RETAIL                      │
│                                                         │
│      Wine and Spirit Merchants, and Brewers.            │
│                                                         │
│   FAMILIES AND THE TRADE SUPPLIED WITH 9, 18 & 36 GALLON CASKS. │
│                                                         │
│        Special attention paid to Private Trade.         │
│                                                         │
└─────────────────────────────────────────────────────────┘
```

Darlaston Brewery.

William Howells, Bilston Street, 1921-23.
Thomas Jones, High Bullen, 1878.
William Lee, Cross Street, 1914.
Arthur Lester, Bull Street, 1914-21.
George Lucas, The Green, 1914-39.
Patrick McCormack, Horton Street, 1914-21.
Joseph Nightingale, Victoria Road, 1914. Mrs Patience Nightingale, 1921.
Samuel Padgett, St John's Road, 1921.
Joseph Ernest Phipps, The Green, 1914. Mrs May Phipps, 1921.
George Roberts, The Green, 1914.
J. & S. Robinson, Horton Street, 1878.
James Rose, King Street, 1878.
Jabez Rubery, Blackmoore's Lane, 1828.
Harry Sheargold, Bull Stake, 1914.
William Shillington, Cock Street, 1828.
John Thomas Simmonds, Bentley Road, 1914-21.
Alfred Small, Walsall Road, 1921-39.
William Stanbridge, Bilston Street, 1914-21.
Richard Turley, Willenhall Street, 1914.
John Wardle, Butcroft, 1914-23. 21, Walsall Road, 1935.
Joseph William Whitehouse, Bilston Street, 1923.
Richard Whitehouse, Cramp Hill, 1914.
Charles Wilkes, The Green, 1868-78.
William Winsper, New Street, 1914.
Thomas Wood, Church Street, 1828.
Henry Woodhall, Cross Street., 1914-23.
George Worrall, Bilston Street, 1914.
John Yates, New Street, 1914-23.

DUDLEY

Albion Inn, 15 Stone Street. A home-brew house in existence by 1830. It was formerly known as the Wellington Arms. It was kept during the 1920s by champion jumper Joe Darby. There is a statue of him in Netherton.

Alma Inn, 91 Hall Street. A home-brew house with brewery to the rear. Originally known as the Travellers Rest, it was established around 1830. The house was renamed the Alma in 1861, in honour of the British victory at Alma in 1854 during the Crimean War. The pub was acquired by Henry and Benjamin Woodhouse in 1901. Brewing apparently ceased in 1914 when the company purchased the Victoria Brewery across the road. Thomas Woodhouse remained as licensee until 1939, when he sold the premises to Julia Hanson & Sons.

Angel, 9 Castle Street. A home-brew house in existence by 1818. Sarah Baker is listed as a brewer and victualler in 1820. The Cole family held the license from the 1830s through to the 1880s. The pub ceased brewing in 1918 when it was acquired by Frederick Smith of Aston in Birmingham.

Apollo Tavern, New Street. Richard Bourne was brewer and victualler here in 1820.

Isaiah Aston, High Street, Kate's Hill, 1914.

David Baker, Salop Street, 1914.

Barley Mow Inn, 36 Constitution Hill. Joseph Billingham was landlord of this home-brew house in 1850. The house was later acquired by Wolverhampton & Dudley Breweries.

Barrel Inn, 173 High Street. First recorded in 1818, with Obadiah Gilbert Shaw as landlord. He was succeeded by his son, Gilbert, in 1819. This home-brew house closed in 1915.

Mrs Ada Baylis, Tower Street, 1914-23. Thomas Baylis, 1930-39.

Enoch Beardmore, 125 Salop Street, 1914-23.

The Beerhouse, Bond Street. Established by retail brewer John Waring following the passing of Wellington's 'Beer Act' of 1830. He is also listed at Oakywell Street.

The Beerhouse, Hall Street. A home-brew house opened by 1846.

The Bell, High Street. First listed in 1820 when Pharez Shore was landlord and brewer.

Belle Vue Inn, 21 Dock Lane. A home-brew house owned in 1850 by Thomas Mansell. It was later bought up by Eley's Stafford Brewery, and closed just before the outbreak of the Second World War.

Bird in Hand, a home-brew house in New Street. John Hobson was landlord in 1820.

Blue Boar, 27 Stone Street. A home-brew house formerly known as the Blue Pig in 1819. William Hughes was licensee. The house closed in 1937.

Blue Gates Inn, 58 Church Street. A home-brew house also known as The Gate, it was in existence by 1818. It was bought by Holden's in 1960.

William John Brecknall, 95 Wolverhampton Street. Established in 1877, this home-brew house was taken over by Eli Bradley of Lower Gornal in 1884. In 1887 the premises were sold to Ansell's of Aston, Birmingham.

Brewer's Arms, 50 Birmingham Street. First listed in 1780 when its landlord and brewer was James Jackson. It remained in the possession of the family for nearly one hundred years. The Brewer's was later bought by Peter Walker around 1900, who in turn sold it to Frederick Smith's Brewery of Aston, Birmingham.

Britannia, 18 Queen's Cross. A home-brew house dating from 1855, it was sold by Emma Woolley to M&B in 1942.

Britannia Inn, 96 Hall Street. Originating in the seventeenth century, this home-brew house was rebuilt in the nineteenth century. It was bought by Frederick Smith of Aston, Birmingham, in 1926. It closed in 1961.

Thomas Broadhurst, St John Street, Kate's Hill, 1923.

John Buckley, Church Street, 1914.

Bull's Head, Hall Street. Originally known as the Old Bull's Head, it dates from pre-1820, when Ezekiel Trowman was listed as licensee. The pub stopped brewing in 1902.

Bull's Head, Stone Street. First recorded in the directory of the town for 1820. William Whitmore was its landlord.

Bush or **Old Bush**, Castle Street/High Street. A home-brew house dating from 1820, when James Cartwright was recorded as its landlord.

William Thomas Butcher, 42 Wolverhampton Street, 1935-39.

California Inn, 13 George Street, Kate's Hill. The house was bought up by W&DB in 1920, whereafter all brewing ceased.

Castle & Falcon Hotel, 207 Wolverhampton Street. Originally the Castle Inn, a home-brew house, with William Bailey as landlord in 1820. At the opening of the twentieth century it was owned by Frederick Smith's of Birmingham. The premises were later sold to the Castle Hotel, 253 Castle Street. The pub ceased brewing in 1947 when it was demolished and rebuilt as a W&DB house.

Thomas Clulow, The Croft, Woodside, 1914-21.

Coach & Horses, 42 King Street, a home-brew house first recorded in 1820 when its landlord was William White. Later owner Arthur Edward Lloyd sold the premises to North Worcestershire Breweries of Stourbridge in 1896. The house was later taken over by W&DB.

Colliers Arms, High Street. Dating from pre-1820, Joseph Morris, victualler and brewer, held the license then. See: Bert & Don Millard in the main listing.

Court House Tavern, 25 New Street. Now a Hanson's pub, this home-brew house was established by 1860.

William Crew, 46 Wolverhampton Street, 1892-1900.

The Cross, High Street. Its first directory-recorded licensee was William Power, in 1820.

Cross Keys, 39 Oakywell Street. A home-brew house, established by 1860, when Edward Sheldon was landlord. Taken over by Hanson's of Dudley in 1900. It closed in April 1957.

Crown & Anchor, 104 Hall Street. A home-brew house dating from 1820 when Thomas Webb was landlord. Later owner and brewer William Oliver Green was fined £10 in August 1887 for attempting to defraud the Inland Revenue by adding 2lbs of sugar to his brew to increase the specific gravity after he had paid duty at a lower rate.

Crown Inn, 29 Crown Street. A home-brew inn whose licensee in 1822 was John Stokes. The pub was bought by W. Butler's brewery and brewing ceased at the Crown. It closed in 1963.

Currier's Arms, Hall Street. A home-brew house dating from 1820, when Joseph Shaw was listed as licensee.

R. Dainty, 37 High Street, Holly Hall, 1896-1900.

Edward Dale, 20 King Street, 1914.

Dragon Inn, 180 Tower Street. A home-brew pub in existence by 1832. It was closed post-1881.

Dudley Arms Hotel, 39 High Street. A home-brew house dating from 1780. It was bought by Netherton brewer Daniel Rolinson, who ran it at a loss. The hotel was acquired by W&DB and closed in 1968.

Duke of Sussex, 78 Stafford Street. Established by 1840 when the landlord was Tommy Tranter. It was bought by W&DB but lost its license in 1916.

Duke of Wellington, 46 Wolverhampton Street. This home-brew house was opened around 1842. It was bought by the Crew family, who had a further brewery behind the Peacock Hotel in Upper High Street. Legendary Black Country brewer Tommy Booth, of Pensnett, was a former brewer and landlord here before it was acquired by Julia Hanson's. The Duke closed during the 1970s.

Duke of York, King Street. S. Wilcox was listed as landlord of this home-brew house in 1818. This pub should not be confused with the one below of the same name.

Duke of York, 128 Wolverhampton Street. A home-brew house dating from pre-1818. John Dickinson was listed as landlord in 1820. The Duke was bought by Hanson's and closed in the 1960s.

Eagle Inn, 14 Dock Lane. Also known as The Eagle, this home-brew pub dates from pre-1850. It was taken over by W&DB in 1923.

Mrs Esther Evans, Lower Dixon's Green, 1914.

Mrs Susannah Fellows, Hall Street, 1921.

Field House Cottage, 43 Oxford Street. A home-brew house in existence by 1870. It ceased brewing in 1940. Acquired by W&DB, it closed in 1953.

George Flavell, The Coppice, 1914-21.

Mrs Alice Fletcher, 46 Wolverhampton Street, 1914.

Edward Fletcher, 27 Brown Street, Kate's Hill, 1935.

James Henry Fones, 148 High Street, 1914-23. David Fones, 1930.

Fox Inn, 502 Wolverhampton Street. This home-brew house was formerly known as the Fox & Goose. It was bought up by Julia Hanson in 1946 and closed down.

Fountain Inn, 3 Dixon's Green. Taken over by M&B in 1920.

Four Ways Inn, 27 Brown Street, Kate's Hill. Taken over by W&DB in 1941.

Freebodies Tavern, 69 St John's Road, Kate's Hill, 1914.

Charles Russell Garton & Co., 272 Castle Street, 1878.

The Gate, Church Street. Robert Garrett, brewer and victualler, was recorded as licensee here in 1820.

George & Dragon Inn, a home-brew house established in the town by 1818, when Eliza Dudley was listed as landlady. By 1820 it was owned by a 'Mr Price', who is described in the directory of that year as a 'victualler & bottled porter dealer'. The George & Dragon closed about 1875.

Golden Fleece, 30 Oakywell Street. A home-brew house in existence by 1870, it was taken over by North Worcestershire Breweries of Stourbridge. It was later sold to W&DB in 1924, when brewing ceased. The Fleece closed in the 1970s.

Cornelius Gorton, Queen's Crescent, 1923.

Green Dragon, Castle Street. Richard Thomas was listed as landlord of this home-brew house in 1820. It was purchased by W&DB and closed in the mid-1960s.

Green Man, Castle Street. A home-brew house in existence by 1820 when Joseph Jewkes was landlord.

Griffin, 8 Stone Street. In existence by 1818, its first known licensee was William Fullward. Rebuilt in 1837, it was taken over by W&DB.

Gypsies Tent, Stafford Street/Stepping Stone Street. Known locally as the 'Tipsy Gent', this home-brew house was formerly known as the Jolly Collier. It was the last home-brew house in Dudley proper. It closed in 1980. The building still remains. See: Bert & Don Millard in the main listings.

Half Moon & Seven Stars, Hall Street. Home-brew house in existence by 1820 when Samuel Paskin held the license.

Mrs Mary Hamblett, Flood Street, 1914.

Hammer Inn, 56 Stafford Street. A home-brew house, William Bullock was landlord here from 1873 to 1901. His popular home-brew was known as 'Bullock's Blood'. The house was acquired by Ansell's in 1938.

Hare & Hounds, (also known as the 'Old Hare & Hounds) 12 Birmingham Street. A home-brew house in existence by 1820, when William Kennedy was its licensee. It was later acquired by Atkinson's of Aston, Birmingham.

Samuel Harper, Oakywell Street, 1914.

William Harper, 1 Brewery Street, Kate's Hill. Established by 1877, the firm was taken over by the Penn Brewery Co. of Wolverhampton in 1897.

Arthur William Hawkins, Stone Street, 1914.

Thomas Hayden, Oakywell Street, 1914-21.

Hearty Good Fellow, Birmingham Street. A home-brew pub originating pre-1820. William Henslow was listed as landlord in that year. Joseph Whitehouse was landlord in 1835. The house closed in 1841.

Hearty Good Fellow, 9 The Square, Woodside, 1938.

Hearty Good Fellow, 8 Flood Street, 1935.

Hollister & Thompson Bros, Stone Street, 1868-72. Hollister was also director of the Handsworth & Perry Barr Brewery of Handsworth near Birmingham.

J.E. Hollows, 60 St John Street, 1908.

Holly Bush, Stone Street. Its licensee in 1820 was Job Fisher.

Holly Bush, Tower Street. The second of three public houses of the same name in Dudley. Its first known licensee was David Shaw.

Holly Bush, Wolverhampton Street. A home-brew house originating pre-1812 with William Collins as brewer and licensee.

Mrs Amy Hopcraft, Church Street, 1921-23.

Horse Shoes, 93 Hall Street. A home-brew pub established in 1820 by Samuel Whyley. Still brewing in 1900, it was acquired by W&DB in 1925. It closed, and was demolished just prior to the Second World War.

The Hotel, High Street. A hotel brewing its own beer for its clientele. In 1820 John Blewitt was listed as licensee.

Heber Hubbold, Pensnett, 1880-88.

George Hughes, 148 High Street, 1935-39.

Walter Humphreys, Dock Lane, 1914-21.

Herbert Hyde, 45 Church Street, 1914.

King William, 9 Cole Street, Darby End. A home-brew house established by 1870. It was run by Daniel Batham, founder of Batham's Brewery, post 1900. The King William was sold to Julia Hanson in 1915, but Batham held the license until 1921. The King William was demolished and rebuilt in 1956.

King's Head. This home-brew house is listed in the trade directory of 1820, when Hannah Johnson was listed as landlady. No address is given. Apparently not the pub of the same name in Birmingham Street, which was run by Susannah Millard in 1820, which has a separate entry.

ISAIAH ASTON

LEOPARD INN,
Kate's Hill, Dudley.

SPARKLING
HOME-BREWED ALES.

FAMILIES SUPPLIED WITH SMALL CASKS.

The Leopard Inn Brewery, Dudley.

Leopard Inn, Church Street. A pre-1820 home-brew house whose licensee in that year was Z. Slaney.

Leopard Inn, 25 High Street, Kate's Hill. Bought up by J.F.C. Jackson in 1926. See: Diamond Brewery.

The Lion, High Street. An early town centre inn and home-brew house. Joseph Gwinnett was listed as licensee in 1820.

George Lloyd, 27 Brown Street, Kate's Hill, 1914-21.

Locomotive, Trindle Road. A home-brew establishment dating from the late 1860s. At the peak of its trade it was run by the Jackson family who produced, 'Jackson's Pure Home Brew'. A sign to this affect was painted across the front of the house. The Locomotive closed in March 1955.

Lord Wellington, See: The Malt Shovel, Tower Street.

John Mark Loverock, 37 High Street, Holly Hall, 1892.

Loving Lamb, 2 High Street, Kate's Hill. Taken over by Darby's Brewery in 1937.

Malt Shovel Inn, 191 High Street/Castle Street. A home-brew house dating from pre-1820. John Mound was listed as landlord in that year,. It closed one hundred years later, in August 1920.

Malt Shovel Inn, Tower Street. A home-brew house originally known as the Lord Wellington, under licensee James White in 1820. He was succeeded by Charles Wilkinson. The house continued brewing up to 1940 and is now a W&DB house.

Henry Mark, Dixon's Green, 1935.

Marquis of Granby, Stone Street. A home-brew house, established pre-1820.

James Mason, 128 Wolverhampton Street, 1914.

Benjamin Mills, Hall Street, 1923.

Richard Mills, Oxford Street, 1914-39.

Miners Arms, 98 Salop Street. Sold to W. Butler & Co. Ltd in 1939.

Mrs Ann May Moore, High Street, Kate's Hill, 1921-23.

Thomas Morris, Flood Street, 1864.

New Cottage Spring, 45 Church Street. A home-brew house built prior to 1850, when its landlord was William Harper. It closed in 1957.

New Inn, 12 Flood Street. Taken over by Julia Hanson in 1922.

Off License, 96 Wolverhampton Street. A retail brewery run by grocer, Thomas Evans during the first decade of the twentieth century. The Off License closed in 1916.

Old Hen & Chickens, Castle Street. A home-brew house in existence by 1820, with Aaron Bennett as licensee.

Old Inn, Wolverhampton Street. Joseph Cardoe was listed as landlord of this home-brew house in 1820.

Old Priory Inn, 15 New Street. Formerly known as the Britannia, this home-brew house was established by 1820 when John Wilkinson was landlord. The change in name occurred after 1850. The house was leased to Ind Coope in 1921, then Holt's of Aston in 1934, who were taken over by Ansell's. The Old Priory later became a Holt, Plant & Deakin house.

Old Struggling Man, 95 Wolverhampton Street. Formerly the Old Inn, a home-brew house in existence by 1866. It was taken over by W&DB and closed in the early 1970s.

Old White Swan, 3 Castle Street/High Street. A home-brew house, also known as the Swan Inn. Its landlord in 1820 was William Pinnock. The Swan was acquired by Hanson's, but was later sold to M&B.

Old Windmill, 13 St James' Terrace. A home-brew pub built on the site of an old windmill and established by 1860. It was bought by W&DB during the 1920s and closed down in 1941.

Parrot Inn, 30 King Street. A home-brew house established by 1873, with Jesse Jewkes as landlord. It closed in 1930.

Peacock Hotel & Brewery, 161 Upper High Street. Its first recorded licensee was Joseph Timmins (or Timmings) in 1820. The hotel was owned at one time by Cllr Solomon Crew, who also owned a number of other pubs in the town. Julia Hanson took over the hotel in 1895. The Peacock closed its doors for the final time in 1927.

George H. Pearson, St James' Terrace, Eve Hill, 1914.

Plume of Feathers, 148 Upper High Street. This home-brew house was established by 1792. It was purchased by George Hughes (grandfather of John Hughes, of the present Sarah Hughes Brewery). The Feathers closed in 1964.

Thomas Poole, Dixon's Green, 1914-30. Mark Henry Pole, 1939.

Richard Price, 141 Holly Hall, 1892-96.

Railway Inn, Bond Street. A home-brew house established by 1830 under Thomas Deeley. It was renamed the Railway Inn with the coming of the railway to Dudley during the 1850s.

Railway Tavern, The Croft, Woodside. Bought by Julia Hanson in 1934.

Red Lion, 31 Bath Street. First listed in the trade directory of 1830, when Samuel Knock was its landlord. It ceased brewing in 1954.

Reindeer, 21 Oakywell Street. A home-brew house originating with William Davies during the 1850s. The family sold the property to Julia Hanson in 1901. The Reindeer closed in 1938.

John Hugh Roberts, St John's Street, Kate's Hill, 1921.

Daniel Robinson, King Street, 1921. Randolph Robinson, 1923.

Roebuck Inn, Stone Street. A home-brew house recorded as early as 1793, when Job Fisher was its landlord. Originally known as The Bush, its name was changed about 1820. It closed sometime shortly after 1860.

Rose & Crown, 71 Spring's Mire, Stourbridge Road. In existence by 1828 under

William Bailey. In 1883 then landlord Joseph Tennant was declared bankrupt with debts of £2,400. Repaying his debts by August 1887, he was back running the Rose & Crown. Beset by money problems, he was fined £5 for attempting to defraud the Inland Revenue by adding sugar to his beer to increase its specific gravity, after he had paid duty at a lower gravity. The house was later sold to W&DB. It closed in April 1929.

Rose & Crown, 52 Withymoor Road. A home-brew pub taken over by Ansell's in 1920.

Round House, 11 Dock Lane. A home-brew house built around 1870, it closed in 1963.

Royal Exchange, 13 Church Street. A home-brew house in existence by 1825, when the licensee was Harriet Pinnock. During the 1920s it was owned by Millward Bros Ltd. The Exchange closed in 1935.

Royal Oak, Salop Street. In existence by 1820, under the licenseeship of Joseph Pitt, this home-brew house was taken over by William Butler & Co. in 1945, whereafter brewing ceased.

Royal Oak, 26 St Martin Hill Street, 1957.

John & Samuel Rudge, High Street Brewery. Brewers and bottled port dealers. The business was established by 1839. Brewing continued post-1847.

Mrs Elizabeth Russ, Oakywell Street, 1914-21.

Mrs Sarah Russon, Church Street, 1921.

Saracen's Head, 18 Stone Street. A home-brew pub first listed in 1808, when Thomas Palmer was licensee. It was run by Julia Hanson's father, John Mantle, from 1835 to 1850. The house stopped brewing in May 1881 and, interestingly, was taken over by Julia Hanson & Sons at the opening of the twentieth century.

Shakespeare's Head, Hall Street. Here by 1820, Mary Timmings was listed as licensee of this home-brew house.

Ship & Rainbow, High Street. Home-brew house well established by 1820 when William Jenns was listed in the trade directory as its landlord.

Shoulder of Mutton, 29 Dixon's Green Road. Taken over by William Butler & Co. in 1939.

Shoulder of Mutton, Union Street. William Dudley was landlord of this home-brew house in 1820.

Charles Simmonds, The Square, Woodside, 1914-21.

W. Simpson, (off-licence & brewer), 96 Wolverhampton Street, Eve's Hill, 1916. Simpson also owned the Albion Tree in Tipton.

Sir Robert Peel, 35 Salop Street. In existence by 1850, it was still brewing under Charles Brightman in 1913. It was bought up by W&DB in 1941 and closed during the 1960s.

Ephraim Smith, St John's Road, Kate's Hill, 1914.

Smith & Co., Hall Street, 1868.

Star & Garter, 71 High Street, Kate's Hill. A home-brew house until about 1880, when it was bought by Kate's Hill Brewery. The pub closed in 1902.

Joseph Stone, 95 Wolverhampton Street, 1914-30.

Struggling Man, Salop Street. A home-brew house dating from 1828, it was acquired by Atkinson's of Aston in Birmingham but later sold to Julia Hanson in 1929.

Talbot Inn, Wolverhampton Street. Originating from pre-1820, its earliest known directory-entered publican was William Beesley.

Mrs Mary Teague, Cross Gun Street, Kate's Hill, 1914.

Three Crowns, Castle Street. A medieval sign evolving from the Three Wise Men, or Kings, this was a home-brew house. Its licensee in 1820 was Mary Beaumont. There was a second Three Crowns in the nearby High Street, whose landlord that same year is given as John Wilkinson.

Arthur Timmins, 136 Salop Street, 1914-23.

Benjamin Tranter, 42 Wolverhampton Street, 1914-30.

Trust in Providence, 7 Washington Street, 1912.

Unicorn Spirits Vaults, 124 Salop Street. Originally called the Somer's Arms, this home-brew pub closed in 1925.

Joseph Upton, Church Street, 1923-30.

Victoria Inn, 23 Dudley Wood. Taken over by William Butler & Co. in 1951.

Vine Inn, 23 Flood Street. Taken over by Julia Hanson in 1901.

Vine Inn, 42 Wolverhampton Street. Bought up by William Butler & Co. in 1951.

Joseph William Wade, 152 Wolverhampton Street, 1914-39.

Wagon & Horses, Hall Street. Joseph Lear was landlord and brewer in 1820.

Wagon & Horses, Spring Mire. A home-brew house. Its landlady in 1820 was a Mrs Harper, possibly the wife or mother of Joseph Harper of the Miners Arms.

James Walters, 56 Stafford Street, 1915.

West End Hotel, 64 Wolverhampton Street. A home-brew house, formerly known as The Boards. It was established by 1850, when it was being run by Mary Ann Wood. Brewing continued here up to 1951, when it was taken over by William Butler & Co. The hotel closed in the 1970s.

Ellen Whitehouse, retail brewer of Bond Street, 1830.

White Swan, 6 Swan Street. This former home-brew house is now a Pardoe's house. Rebuilt in the 1970s, the original house dated from 1865. It was briefly owned by Atkinson's of Aston during the 1920s and 1930s.

Roger Whitmore, 27 Brown Street, Kate's Hill, 1923-30.

Why Not Inn, 23 Abberley Street. A home-brew house opened in the early nineteenth century. It was bought up by Julia Hanson in 1946. The name was later changed to The Fox.

William Wilkinson, 15 New Street, 1914.

Joseph Willetts, St James' Terrace, 1921-23.

Thomas William Williams, Tippity Green, 1908-24.

Edward Winby, Queen's Cross, 1921.

Windmill, Wolverhampton Street. Joseph Round is the earliest recorded landlord of this home-brew house. He is listed in the directory of 1820.

Mrs Mary Witcomb, Queen's Cross, 1921.

Wonder Inn, 52 Church Street. A home-brew house formerly called The Crown. It was bought by Hanson's in 1928, whereafter brewing ceased.

Joseph Wood, 64 Wolverhampton Street, 1914.

Benjamin Woodhouse, Hall Street, 1914-21.

T. & B. Woodhouse, Queen's Cross, 1923-30.

Woolpack, High Street. George Glazebrook was listed in 1820 as owner of this home-brew house.

Woolpack, 15 Castle Street, 1960.

Edward Wright, Dock Lane, 1914-35.

John Walter Yates, Bath Street, 1914-23. At Church Street in 1935. Mrs Janet Mary Yates in 1939.

ENVILLE

James Dorrell, 1892

GORNALL WOOD

Isaac Bradley, 1921-23. Premises sold to the Holt Brewery of Aston, Birmingham, in 1930.

Richard Bradley, retail brewer, listed in *Bridgen's Directory of Wolverhampton and Neighbourhood* for 1833.

Edward Guest, 1833. Also a gardener.

James Hughes, retail brewer and butcher, 1833.

Old Bull's Head, Redhall Road. This home-brew house was in existence by 1834, when Edward Guest was listed as its licensee. The brew house at the rear still exists and is a listed building.

HALESOWEN

Albert William Beale, Hasbury, 1914-20.

Mrs Teresa Cartwright, Gorsty Hill, 1914-20. William Cartwright, Gorsty Hill, 1930.

Joseph Cattell, Quinton, 1880. [Quinton is now part of Birmingham].

Mrs Harriet Cooke, Long Lane, 1896.

Tom Cresswell, Peckingham Street, 1923-24.

Benjamin Darby, Mucklows Hill, 1900.

George Grainger, Birmingham Street, 1914-21.

W. Grainger, Little Cornbow, 1900-08.

Mrs Ann Green, Birmingham Road, 1884-96.

Mrs Lydia Hackett, Islington, 1914-20.

John D. Harris, High Street, 1914-30.

Hodgetts & Cooper, Church Street, 1884-92.

Ernest John Hollies, Peckingham Street, 1945.

Harry Hollis, Forge Lane, 1914. Mrs Harry Hollis, Forge Lane, 1921.

Edward Jackson, Birmingham Street, 1923.

Samuel Jones, Great Cornbow, 1914.

Edward Lowe, Gorsty Hill, 1914-23.

Samuel Lowe, Gorsty Hill, 1914-26.

Alice Marshall, Birmingham Street, 1926.

Mrs Alice Marson, Peckingham, 1930.

Joseph James Moore, Birmingham Street, 1914. Edith Moore, Birmingham Street, 1921.

James Mynett, Great Cornbow, 1914.

New Road Brewery, J. Bloomer & Sons, brewers, 1910.

Paskin & Co., 1880.

Mrs Hannah Rollason, Cocksheds, 1914. Cockhead Lane, 1921.

William Shuker, New Street, Hasbury, 1914-20.

George Smith, Birmingham Street, 1914-20.

Edward Sturman, Long Lane, 1914-20. Also a wine and spirit merchant.
Percy Withers, Birmingham Street, 1923-40.

KINGSWINFORD

Arthur Thomas Allen, 13 High Street, 1914-35.
Eli Bird, Wallheath, 1914.
Joseph Brettle, High Street, 1923-40.
Richard Chambers, Market Street, 1914-40.
William Joseph Cook, High Street, 1884-92.
Mrs Sarah Edwards, 1914-21.
Elephant & Castle, Bromley Lane. A home-brew house dating from 1838. Taken over
   by Truman, Hanbury Buxton & Co. in 1902. Later acquired by Courage, who sold it to
   Holden's in 1983.
Sydney Marsden, Dawley Brook, 1930.
Daniel Henry Marsh, High Street, 1914-21.
William Marsh, Greens Forge, 1921-21. Mrs Minnie Marsh, Green's Forge, 1923.
Mount Pleasant Brewery, William H. Morgan, 1910.
Henry Parfitt, Common Side, Pensnett, 1923.
Mrs Florence Mary Penn, Cot Lane, 1923-30.
Harry Westwood, Dawley Brook, 1921-23.
William H. White, Summer Street, 1914-30.
John Bright Willis, Bromley, Pensnett, 1914.

KINVER

Felix Broughton, 1921-23.
Dennis Gardener, 1914.
Joseph Gatti, 1914. Mrs Susannah Gatti, 1921-40.
Benjamin Hadley, 1921-23.
Mrs Esther E. Keeley, 1930.
Hardy Lumb, 1921.
Eli Millward, 1930-35.
Mrs Elizabeth Price, 1914-23.
Leonard H. Turner, 1923-26.
John Wrigley, 1914.
John Henry Yates, 1926.

LANGLEY

Arden Grove Brewery Co. Established by S.N. Thompson & Webb in 1877. By 1888 the
   brewery was being run by William Webb. By 1887 it was in the hands of Charles King,
   who sold it to J. Nunnery & Co. in 1893. The brewery was later acquired by Ansell's.
Brewery Inn, Station Road. A canalside pub which brewed Holt's beer on site.

Captain Rose, Causeway, Green Road, 1896.
David Millerchip Sadler, Dog Kennel Brewery, Langley Green, 1884.

LOWER GORNAL

William Henry Anderson, Sheepcote Brewery, 1884.
Thomas Bailey, 1926.
Thomas Bate, Ruiton Street, 1914-30
John Bir, Abbey Road, 1914.
Samuel Bracknell, 1914-26.
Isaac Bradley, Reddall Road. The business was instituted by Eli Bradley in 1887. He was
    succeeded by Emanuel Bradley in 1895, Daniel Bradley in 1898 and was eventually sold
    to the Holt Brewery of Aston by Isaac Bradley in 1930.
Chapel House Inn, Ruiton Street. Owned by William Cartwright, he left the property
    to his niece in his will. She and her husband sold it to James Hughes, who sold it in turn
    to brewer John Kimberley for £410. It became the Miners Arms under brewer Charles
    Evans, who later sold it to Thomas Booth.
Benjamin Evans, 1914-21.
Benjamin Evans jr, 1914.
Elizabeth Evans, 1923-26.
John Evans, 1923-26.
Arthur Fieldhouse, Ruiton Street, 1935.
David Hyde, Lake Street, 1914-30.
John Jones, New Street, 1914-23.
Mrs Matilda Jones, New Street, 1914-23. Eli Jones, 1930-35.
Joseph Jukes, 1914-20.
Thomas Malpas, Abbey Road, 1914-26.
Mrs Emily Marsh, Himley Road, 1914-35.
Moses Marsh, 1914-20. Brewer and wine and spirit merchant.
Thomas Oakley, retail brewer, 1833.
Old Mill Inn, Windmill Street, Ruiton. Established as a beerhouse by John Hyde in 1852
    with a small brewery at the rear. It was bought at auction by Henry Rolinson in 1875.
    An inventory of the pub and brewery, drawn up in 1898, records that the brew-house
    contained '3 Loose Cooling Vats, Loose mash tub, 2 loose Wort Spouts, Mash Rule,
    2 Tubs and Copper and platform, Hop Press Sieve, 2 buckets, loose refrigerator'. After
    various owners, it was bought by Frederick Smith's Brewery of Aston, Birmingham. It
    came into the possession of Holden's of Woodsetton in 1946.
Henry Payton, 1914. Mrs Henry Payton, 1921-23.
Enoch Smart, 1914-30.
Joseph Tomlinson, 1914-23.
Harry Turner, 1914.
William Wakelam, 1930.
John Waterfield, Five Ways, 1914-20.
Joseph Waterfield, 1872-84.
David White, 1914.
William Willetts, 4 Church Street, 1923.

DAVID M. SADLER,

✦BREWER OF FINE ALES.✦

*T*HESE Ales have, in a very short period, gained great favour with the Public; being brewed with the Celebrated Water drawn from the famous Crosswell's Road Springs, and the best Malt and Hops to be procured in the markets.

Dog Kennel Brewery, **LANGLEY GREEN**, near **OLDBURY**,

BIRMINGHAM.

N.B.—See Analysis of the Water, which can be had on application, either personally or by post.

Dog Kennel Brewery, Langley.

Thomas Bailey's Old Bull's Head home-brew house, Lower Gornal.

Evans' home-brew house, The Fountain, Lower Gornal.

THE FOUNTAIN

MOXLEY

Joseph Maloney, Woods Bank, 1921.
Fred Wheale, Queen Street, 1921.

NETHER END

Henry Herring, 1914-35
George Albert Ingley, 1921.
George Johnson, 1923-26.

NETHERTON

Samuel Bagley, Cole Street, Darby End, 1892-96.

Barley Mow, Blackbrook Road. A home-brew house around 1855-70.

Daniel Batham, King William, Cole Street, Darby End, 1914. See: Batham's Brewery.

Francis Henry Billingham, 47 Simms Lane, 1914-39.

Bird in Hand, 82 Chapel Street. An early nineteenth-century home-brew house owned during the 1890s by William Onslow. The Primrose Hill Brewery was built behind it. The pub was later sold to North Worcestershire Breweries of Stourbridge in 1896, while the small brewery was sold to Elijah Bywater, who owned the Colliers Arms in Chapel Street. The pub and brewery were later acquired by W&DB. They rebuilt the pub and brewed here until the early 1970s.

Blue Bell, 15 Cradley Road. A late eighteenth-century home-brew house in existence by 1781, when its landlady was Sarah Rolinson, matriarch of a brewing dynasty. Brewing continued at the Blue Bell up to 1930, under Thomas Ernest Harris. Brewing finally ceased in May 1939.

Thomas Booth, St Andrew Street, 1923-30. See: Thomas Booth's Brewery in the main section.

John Ernest Churchill Bridgwater, 49 John Street, 1914.

British Oak, 91 Sweet Turf. A home-brew house established pre-1861 under brewer Charles Cartwright. The house was later leased to Michael Hodgkins during the latter decades of the nineteenth century. The Oak's last brewer was Edward Prince in 1930. Its last tenants were Frederick and Doris Pardoe, who later moved to the Old Swan. The British Oak was sold by Prince's widow, Louisa, to Julia Hanson in April 1932.

Bull's Head, 19 St John's Street. First recorded in 1828 as a home-brew house. Briefly part of the ill-fated Dudley & District Breweries Ltd, founded in 1896, the house and its outbuildings were acquired by the Netherton Bottling Co. Ltd in 1926. The firm also owned the Railway Tavern in St Michael's Street, West Bromwich. The Bull's Head ceased brewing when it was taken over by W&DB.

Arthur Burchill, St James' Street, 1921-30.

Colliers Arms, 62 Chapel Street. A home-brew house brewing up to 1936, with Elijah, and later his son, Thomas Bywater. The family also owned the Primrose Hill Brewery and the Star in Cradley.

Joseph Davies, Castle Street, 1914. Wheelwright Arms Brewery, Griffin Street, 1939.

Map of Netherton, 1936, showing its home-brew pubs.

Davies had thirteen pubs and an off-licence. His enterprise was taken over by M&B in June 1942.

Herbert Dunn, St Andrew Street, 1921.

John Dunn, Withymoor Road, 1914. Cradley Road, 1923.

Five Ways Inn, St Andrew Street. A home-brew house established by 1832 under landlord Thomas Penbury. It developed into the Five Ways Brewery under John Rolinson.

Golden Lion, 5 Simms Lane. A home-brew house under Joseph Smart, who sold it to Ansell's in 1920.

George Garratt, St Andrew Street, 1914.

William Griffin, 62 Chapel Street, 1914-21.

William Griffin, Cradley Road, 1921.

Thomas Ernest Harris, Blue Bell, 15 Cradley Road. This home-brew house was established by 1876. Harris was brewing here until 1930. The Blue Bell closed in May 1939.

Thomas Hotchkiss, Castle Inn, High Street. Hotchkiss, a miner and victualler, was licensee in 1835. After his death his sons, Thomas and William, continued the business. They styled their enterprise 'The Old Brewery, Ale & Porter Brewers, Sweet Turf'. They appear to have ceased brewing sometime after 1892.

Loyal Washington, Washington Street. A home-brew house and street named after William Washington, who had built a pub here, *c*. 1863. The house was rebuilt in 1901 when it was taken over by Plants. It was later acquired by Ansell's, just prior to the outbreak of the Second World War.

Albert Lyndon, Halesowen Road, 1914-23.

Netherton Brewery, Cinder Path. Founded in 1852 by William Woodward Smith, maltster and brewer. He also owned the nearby Hope Tavern between 1864 and 1884. Smith briefly operated the Dudley Brewery in Hall Street, as William Smith & Co.

Netherton New Brewery was situated behind the Old Pack Horse, 39 Hill Street. It was founded by Richard Rolinson and opened in 1875. His son, John Rolinson, was put in charge of the firm, which became John Rolinson & Son of St Andrew Street in 1891. The brewery was almost destroyed by fire on 26 March 1895. The following January John Rolinson died at the age of seventy-four. The brewery later amalgamated with the Five Ways Brewery in 1900. The company, listed as John Robinson & Son Ltd was taken over by W&DB. William Onslow, formerly of the Bird in Hand, brewed at the Pack Horse, the brewery tap, from 1898.

MARK FLETCHER,

LOYAL WASHINGTON INN,

NETHERTON, near DUDLEY.

Ales, Wines, Spirits, &c., of the best Quality.

Loyal Washington, a home-brew pub.

New Inn, 31 High Street. A home-brew house built prior to 1870, when Samuel Taylor was licensee. It was advertised for sale in June 1901 and included a 'sixteen bushel mash tun, two hundred gallon steam boiler and five fermenting vessels'. The equipment was capable of producing thirty-six barrels of beer at a time. The house was bought by North Worcestershire Breweries of Stourbridge and the brewery was closed down. The pub was later acquired by W&DB.

Old Cottage Inn, 24 Simms Lane. A pre-1870 home-brew pub owned for a number of years by the Hampton family. It was situated alongside the Netherton New Brewery. Home brewing continued at the Cottage up to the outbreak of the Second World War.

Old White Swan, 6 Halesowen Road. Established by 1835 when Thomas James was licensee. Rebuilt in 1863, it was affectionately known as Ma Pardoe's. See the main listing.

Primrose Hill Brewery, 82 Chapel Street, rear of the Bird in Hand. William Onslow and his family brewed here from before 1864 to 1901. The brewery was later bought by Elijah Bywater, who also owned the nearby Colliers Arms in Chapel Street and The Star at 45 Cradley Road. Brewing continued with Elijah's son, Thomas, up to 1936.

Edward R. Prince, Union Street, 1923-30.

Queen's Head, 45 St John's Street. A home-brew house established by 1875, when it was owned by the Billingham family, who retained ownership for over seventy-five years. Brewing ceased in 1935.

Francis Ratcliffe, 29 Hill Street, 1904.

Reindeer, 16 Cradley Road. This home-brew house was owned by the Tibbetts family during the mid-nineteenth century. It was leased to Butler's in 1929, when brewing ceased, and was finally closed in 1938.

Mrs Mary Ann Roe, Baptist End, 1914-39.

Rose & Crown, 52 Withymoor Road. Mrs Louisa Bird, 1892-96. A home-brew house. William Dunn bought the pub in 1897 and it passed on to his son, John, in 1920.

Samuel Round, High Street. Offices in St John Street, 1864-72. Thomas Round, 1914-23.

49 St John Street. A retail brewery owned by Benjamin Bridgewater in the early twentieth century. It was taken over by the Netherton Bottling Co. in 1926.

Joseph Smart, Golden Lion, 5 Simms Lane. A late nineteenth-century retail brewer fined £10 for defrauding the Inland Revenue by adding sugar to his brew, thus increasing the gravity beyond the duty paid. He continued to brew after this, selling up to Ansell's in 1920.

The Star, a home-brew house. In August 1887 brewer George Robert Chatham was fined £20 for adding sugar to his brew to increase the specific gravity, thereby defrauding the Inland Revenue.

William Tibbetts, Cradley Road, 1914.

Thomas Tromans, Castle Street, 1921-35.

Trust in Providence, 7 Washington Street. A home-brew house built before 1840, along a pathway that later became Washington Street. At one time it was owned by Benjamin Round, whose daughter, Lucy Blanche, married Edwin Holden of Holden's Brewery in October 1898. The Trust closed in December 1912.

White Horse, 62 St Thomas Street. A home-brew house established by 1870 under landlord William Burchill. It was taken over by Ansell's in 1938, when brewing ceased.

SAMUEL·ROUND,
BREWER OF CELEBRATED
WORCESTERSHIRE, PALE, MILD, AND STRONG
ALES,
And DOUBLE BROWN STOUT.
THE BREWERY,
Office: ST. JOHN'S STREET, NETHERTON, near DUDLEY.

COTTAGE SPRING BREWERY,
NETHERTON, WORCESTERSHIRE.
SAMUEL ROUND,
PALE, MILD, & STRONG ALES & PORTER
BREWER,
Maltster & Hop Factor. Wine & Spirit Merchant.
HIGH STREET (OFFICES, ST. JOHN STREET),
NETHERTON, DUDLEY, WORCESTERSHIRE.

Above: *Advertisements for Round's Cottage Spring Brewery.*

Right: *Swain's Albion Brewery, Oldbury.*

White Swan, 23 Baptist End. A home-brew house established by 1830, when Sylvia Smith was recorded as landlady. In 1939, then landlady Mary Ann Roe sold the property to Ansell's.

NEW INVENTION

John Brooks, 1914-21.
Moses Parkes, 1878.

OLDBURY

Albion Brewery, Tatbank Road. Henry Swain & Co. 1921.
Arthur Astley, 19 Freeth Street, 1896-1908.
Bath Row Brewery, William Hadley, 1913.
Arthur Duffield, 30 Flash Road, 1896.
John Evans, 74 Birchfield Lane, 1896-1900. Joseph Evans, Birchfield Lane, 1904-21.
M. Gibbs, 84 Birmingham Road, 1896-1908.
Mrs Hannah Hadley, 10 Rounds Green, 1904-26.

William Hadley, Bath Row Brewery, Rounds Green, 1888-1912. Hadley died in 1913 and his brewery was run by executors from 1914-20.

J. Hartland, Dudley Road, Brades, 1896.

Henry Jerrams, Parsonage Street, 1878-86. Golden Ball Brewery, 1888. George Jerrams, 1892.

John Jordan, Queen's Brewery, 44 Birmingham Road, 1880. In 1884 the directory refers to it as the British Queen Brewery, at the same address. In 1888 it is the British Queen Brewery (No.1) 44 Birmingham Road; (No.2) Langley Green, 1892-1914. John Jordan & Co., British Queen Breweries, 1916.

James Lines, 6 Flash Road, 1896-1912.

Walter Middleton, 150 & 152 Birchfield Lane, 1900-08. Mrs Maria Middleton, Birchfield Lane, 1914-21.

William Nickliss, 2 Rounds Green, 1896-1900

David Noake, 49 Halesowen Road, 1896.

Joseph Page, 14 Green Street, 1896-1904.

Henry George Peers, Talbot Street, 1914. Henry George Peers jr, 7 Talbot Street, 1921-35.

William Round, Park Street, 1914-21.

Nathaniel Sadler, Windsor Brewery, Dingle Street, Rounds Green, 1904-28.

Thomas A. Sadler, Dingle Street, Rounds Green, 1923-26.

Samuel Sherwood, 11 Church Bridge, 1896 1940.

Harry Snead, 37 Bath Row, 1896.

Henry Swain, Albion Brewery, Tat Bank Road, 1880-1912. Henry Swain & Co., 1914.

Stephen Turner, 15 Talbot Street, 1896.

John Ward, 49 Tat Bank Road, 1904-14. Frank Ward, Tat Bank Road, 1921.

George Williams, Chapel Street, Round's Green, 1926.

John Yorke, Birmingham Street, 1921-23.

OLD HILL

Alfred Cockin, Cherry Orchard, 1914-30

Arthur C. Cockin, Halesowen Road, 1921-30.

Thomas Darby, 61 & 64 High Street, 1904.

Thomas Edge, Halesowen Road, 1914.

John Foley, 1878-88.

William Foley, Halesowen Road, 1914. Henry E. Foley, 1921-23.

Albert Ernest Hadley, Reddall Hill Road, 1914. Henry E. Hadley, 1920.

Ernest Hall, Cradley Road, 1914-20.

Thomas James, Halesowen Road, 1914-40.

Arthur Ernest Jew, 77 High Street, 1935-40.

George Johnson, Halesowen Road, 1914-40.

Daniel Perry, Reddall Hill Road, 1914. Samuel Perry, 1921-23. Alfred Perry, 1930. Mrs

Laura Perry, 128 Reddall Hill Road, 1935-39.

Thomas Priest, 128 Waterfall Lane, 1908.

Albert Eland Sidaway, 100 High Street, 1904-23. A.E. Sidaway & Son (exors) 27 High Street, 1924-30.

James Stafford, Ridding Street, 1914-20.
Henry Swain, Albion Brewery, Tat Bank Road, 1884.
Thomas William, Cherry Orchard, 1935-40.

PENSNETT

George Dunn, Bell Street, 1914-20.
John Fradgley, Church Street, 1921.
Laban Hill, Queen Street, 1914-21.
Arthur Jones, Church Street, 1914.
James Henry Parfitt, Common Side, 1914-30.
John Bright Willis, Bromley, 1914.

QUARRY BANK

Roland Batham, Elephant & Castle, High Street, 1919-45. Roland was a brother of
   Daniel Batham. Batham's bought the house in 1940, but Roland retained the license until
   retirement in 1945.
Home Brewery, William Thomas Clewes, brewer. Clewes owned the Red Lion in
   Gornal Wood, with William Jones as his licensee.
Samuel Mobberley, Sheffield Street, 1914. Nock & Co., 1915-40.
Mrs Matilda Nock, East Street, 1914. Nock & Co., 1915-40.
Henry Stevens, High Street, 1914-30. Clara Stevens, 166 High Street, 1935.

ROSEVILLE

Albert Asford, Castle Street, 1930.
John Arthur Grange, Castle Street, 1921-40.
John Horton, Castle Street, 1921-23.
Mrs Ruth Millard, Ward Street, 1914-26.
William Millard, 1914-20.
David Pearson, Castle Street, 1914.
Thomas Porter, Ebenezer Street, 1914. Eder Porter, Ebenezer Street, 1921-30.
James Richards, Ward Street, 1914-40.
Samuel Timmins, Ward Street, 1914-40.
James Henry Yates, Castle Street, 1935.

ROUND OAK

Mrs Annie Amphlett, 1923-30.
Herbert Bissell, 1914-21.
Benjamin Elwell, 1892. See: Town Brewery in main list.
Charles Frederick Moore, Wallows Street, 1923-29. Harry Clifton Moore, 1930-40.

Mrs Sarah Preston, Wallows Street, 1914–21.
William Henry Simpkiss, 1884–92. See: Simpkiss in main listing.
Smith & Williams, Town Brewery, 1921–32.
Alfred Tandy, 1914–21.

ROWLEY REGIS

Meshack Hacket, Cocks Green, 1914–21.
John Haden, Rowley Village, 1914–21
Frederick Mason, Springfield, 1914–23.
Joseph Tibbetts, Springfield, 1914.
The Vine, a home-brew house. In 1887 its then landlord, Joseph Walters, was fined £10
   for attempting to defraud the Inland Revenue by adding sugar to his brew to increase the
   specific gravity, after having paid duty at a lower rate.
Thomas William Williams, Tippity Green, 1904–23.

SEDGLEY

Benjamin Bailey, Bilston Row, 1833.
Henry Baker, Skidmore Row, 1833.
Thomas Beaman, Broad Lanes, 1833.
John Beardmore, The Coppice, 1833.
Cutlack Brothers, Sedgley Brewery, 1888. William Cutlack jr, post-1891.
Joseph Dabbs, Momble Square, 1833.
Howard Darby, 1914–21.
Thomas Downing, Prince's End, 1833.
William Dunton, Brierley 1833.
Samuel Elwell, Sodom, 1833.
Thomas Elwell, Fullard's End, 1833
Francis Evans, 1914. Mary Evans, 1921.
Hugh Richard Fellows, High Street, 1914–20.
James Fellows, Bilston Street, 1914–21. Co-owner of the Beacon Hotel. See: Sarah
   Hughes.
John Gittins, Gospel End, 1833.
Edward Glover, Gospel End, 1833. Also a shoemaker.
Eliza Gregory, Bateman's End, 1833.
Ann Grinsell, Daisy Bank, 1833.
James Halford, Hall Green, 1833. Retail brewer and pattern maker.
David Hartland, Sodom, 1833. Retail brewer.
Thomas Hazeldine, Ettingshall Lane, 1833. Retail brewer.
Jonah Hickman, Straits Green, 1833.
John Hickman, Straits Green, 1833.
Joseph Hickman, Gospel Oak, 1833.
Richard Hilton, Can Lane, 1833.
William Hilton, Straits Green, 1833.

John Howell, Broad Lanes, 1833.
Thomas Jackson, Sodom, 1833.
Abraham Jeavons, Hall Green, 1833.
Simeon Jeavons, Fullard's End, 1833.
Mrs S. Jenkins, High Street, 1872.
Francis Jones, The Straits, 1833.
Thomas Kennedy, Princes End, 1833.
John Kimberley, Dudley Road, 1878-84.
John Langley, Mount Pleasant, Fullard's End, 1833.
John Lester, Gospel End, 1833.
John Lewis, Bateman's End, 1833.
Thomas Lisle, Gospel End Street, 1914-23.
Mrs Mary Marsh, Gate Street, 1914-30. Richard Marsh, 1935. Mrs Ada P. Marsh, Gate Street, 1939-40.
Mary Martin, Skidmore's Row, 1833.
Joseph Morton, Prince's End, 1833.
Harry Palmer, 47 Bilston Street, 1914-20.
Philip Partridge, Hall Green, 1833.
James Prince, Fullard's End, 1833.
William Henry Rollinson, 51 High Street, 1914-20.
James Round, Walbrook, 1833.
Benjamin Salt, Can Lane, 1833.
Stephen Sheldon, Deepfield, 1833.
John Shorthouse, Can Lane, 1833.
John Stanford, Fullard's End, 1833.
John Taft, Fullard's End, 1833. Retail brewer and glazier.
Ellen Wallens, Gospel End Street, 1914-20.

SMETHWICK

Henry Bunn, farmer and retail brewer, Wigmore, 1845.
Creswell (Smethwick) Ltd. Founded in 1960.
John William James Kingstone, Summit Brewery, Great Arthur Street, 1921-23.
Edward Lewis, 296 High Street, 1914-26.
Thomas Robinson, publican and maltster, Talbot Inn, a home-brew house, Oldbury Road, 1842-55.
John Rudge, Commercial Inn, Oldbury Road, 1839-47.
Charles Wheeler, 375 Oldbury Road, 1914-21.

STOURBRIDGE

Mrs Caroline Asbury, 19 Enville Street, 1914-22. Eliza Asbury, 19 Enville Street, 1923.
James Samuel Asbury, 77 Brettell Lane, Amblecote, 1920.
Askew & Co., Duke Street and Birmingham Street, 1910.
James Charles Attwood, Birmingham Street, 1900. Stourbridge Road, Lye, 1914. James

Attwood, Upper High Street, Lye, 1914-21. James Charles Attwood, Stourbridge Road, Lye, 1920.

John Auden, Enville Street, 1914-23.

Aubrey O. Bache, Mamble Road, Wollaston, 1923-26.

Mrs Ellen Bache, Cherry Street, 1914. William J. Cache, Cherry Street, 1921.

Arthur Barlow, Old Swinford, 1914.

Frederick Barlow, Mount Street, 1923-30.

Joseph Bate, Wood Street, Wollaston, 1921.

Edward Bennett, 6 High Street, 1914-21.

Mrs Maria Bellingham, Dudley Road, Lye. 1914.

Bird in Hand, 149 Hagley Road, Oldswinford. A home-brew house with brewery at the rear. Originating as the Lion Inn, Job Coley was landlord in 1832. It was renamed The Old House at Home by 1860 and renamed the Bird in Hand by 1896, when it was rebuilt. Daniel Batham bought the pub in 1926, whereafter it ceased brewing.

E.M. Bowker, Wood Street, Wollaston, 1914.

William Bridgewater, Stourbridge Road, Lye, 1914. John Bridgewater, 51 Stourbridge Road, Lye, 1921-26.

Francis Brooks, Worcester Road, West Hagley, 1935.

Jeremiah Brooks, Talbot Road, Lye, 1914-20.

Joseph Brooks, New Road, 1914-20.

Mrs Mary Ann Brooks, The Cross, High Street, Lye, 1914-21. Pedmore Road, Lye, 1930.

Thomas Bywater, King William Street, Amblecote, 1930.

Thomas Cartwright, Talbot Street, Lye, 1921.

Arthur Catlin, New Street, 1914.

Alfred Harry Cook, Upper High Street, Lye, 1921-40.

Alfred Henry Cook, Talbot Street, Lye, 1939-40.

Esther Cook, Bromley Road, Lye, 1910.

James Cook, Pedmore Road, Lye, 1914-25.

John Cook, Careless Green, Lye, 1914-21.

Samuel Cook, Careless Green, Lye, 1920-39.

George Edward Cox, Heath Lane, Old Swinford, 1921.

James Henry Cox, Stourbridge Road, Lye, 1914-20.

Mrs Mary Ann Cox, Heath Lane, Old Swinford, 1914.

Richard Cox, 56-7 Brettell Lane, Amblecote, 1920.

Cross Walks Brewery, Cross Walk, Lye. Mrs L. Penn, 1956.

Charles Croxton, 14 Worcester Road, 1921. Heath Lane, Old Swinford, 1923.

Albert Sydney Danks, Old Swinford, 1923.

Isaac T. Digger, 40 Enville Street, 1914-21.

Henry Edwards, 37 Coventry Street, 1921.

David Evans, Old Swinford, 1921-35.

Samuel Evans, Old Swinford, 1914. Henrietta Evans, Hagley Road, Old Swinford, 1939-40.

Edward Fletcher, 37 Coventry Street, 1914. High Street, implicate, 1926.

Isaac Fletcher, Audnam, 1926.

Jeremiah Guest, Upper High Street, Lye, 1914-20.

Daniel W. Hale, 49 Church Street, 1921-23.

William Harper, Penn Brewery. Later sold to Ansell's.

Haskew, Whitwell & Hand, Duke Street, 1884–96.

Frederick Higgs, 24 Lion Street, 1914.

James Higgs, 35 Coventry Street, 1914. Mrs Mary Higgs, 1921–30. James Higgs, 1935–40.

Alfred E. Hill, Old Swinford, 1914.

Frederick Hill, The Ridge, Wollaston, 1914–21. Benjamin Alfred Hill, The Ridge, Wollaston, 1923–35.

Rowland Hill, Coventry Street, 1923.

Albert Hipkins, Queen's Brewery, Enville Street, 1930–32.

Albert Hipkiss, Mamble Road, Wollaston, 1914–21.

Frederick Hobson, Belmont Road, Lye, 1914–21.

Mrs Isabella Hodgetts, Careless Green, Lye, 1914–21.

William Hughes, 58 High Street, 1914–26.

Mrs Annie Jackson, Stourbridge Road, Lye, 1921.

John Jacobs, Old Swinford, 1914.

James Jeffries, Hagley Road, Old Swinford, 1930–40.

George Johnson, High Street, Lye, 1926.

Mrs Jesse Lashford, 24 Lion Street, 1921–26.

Bert Lewis, Worcester Street, 1914.

William Maiden, Shepherds Brook, Lye, 1914–21.

Frank J. Matthew, Church Street, 1914–40.

William H. Maybury, Orchard Lane, Lye, 1914. Mrs Esther Maybury, Orchard Lane, Lye, 1921.

Bert Middleton, Stamber Mill, 1939–40.

William Miles, 27 & 49 Enville Street, 1914–40.

Mrs Sarah Ann Millward, High Street, Lye, 1914–35. David Millward, 168 High Street, 1939–40.

Mrs Henrietta Moorcroft, 107 Worcester Street, 1914–40.

Levi Newey, Balds Lane, Lye, 1914–30. Thomas Newey, 2 Balds Lane, 1935–40.

Herbert Newnam, Pedmore Road, Lye, 1914. Retail brewer. See: Herbert Newnam & Sons in main listing.

Edward Onions, Birmingham Road, 1921.

Thomas Pagett, 27 High Street, 1914–23.

Frederick Pardoe, Orchard Lane, Lye, 1914. Stourbridge Road, Lye, 1921.

Thomas Pardoe, 113 Stourbridge Road, Lye, 1930–35.

John L. Pargeter, Old Swinford, 1923.

Alfred Parkes, Heath, 1914–21. Emily Parkes, Heath, 1923.

Caleb Parrish, Skeldings Lane, Lye, 1914–21.

Joseph Pearshouse, Hayes Lane, Lye, 1914–21.

John & Horace Pearson, 36 Bridgnorth Road, 1914–21 (in partnership). Horace Pearson, 36 Bridgnorth Road, 1921. John Pearson, Bridgnorth Road, 1921. Two separate companies.

John Penn, Cross Walk, Lye, 1914. Mrs Laura Penn, 1921–28. John Penn, Cross Walks Brewery, Lye, 1930–39.

Mrs Isabelle Perrins, Hagley Road, Old Swinford, 1914–21.

Henry Perry, Green Lane, 1914. John Henry Perry, Green Lane, Lye, 1921.

John Price, Stourbridge Road, Lye, 1914.

Queen's Brewery, Enville Street. Albert Hipkiss, 1933.

Thomas Rhodes, Pedmore Road, Lye, 1914. John N. Rhodes, Pedmore Road, Lye, 1921.

Thomas Frederick Rhodes, 13 Orchard Lane, Lye, 1930-35.

Mrs Emily Robinson, Upper High Street, Lye, 1914.

Mrs Louisa Roper, Angel Street, 1914-23. John Thomas Roper, Angel Street, 1926.

Royal Exchange, 75 Enville Street. A home-brew house dating from 1855, under butcher-turned-publican John Tauberville. Owned by Frank Matthews from 1922 to 1940. His widow Sophia leased the premises to Batham's in 1946.

Edward Rutland, Queen's Head Brewery, Enville Street, with offices in the High Street, 1914-23. The firm became Edward Rutland & Son in 1924. The brewery was taken over by Albert Hipkins in 1929. The Queen's Head Brewery was later demolished, but the brewery tap, the Queen's Head, survives. The Plough and Harrow in Kinver High Street, a former Rutland's house, was later taken over by Batham's.

Mrs Susan Shuker, High Street, 1923.

Harry Skelding, Hagley Road, Old Swinford, 1908-14.

M. Skelding, New Street, 1921-30.

Mrs Mary L. Smith, Cross Walk, Lye, 1914.

Thomas Sparrow, 91 Swinford, 1923-30.

Joseph Taylor, Talbot Street, Lye, 1914.

John Henry Thornes, Old Swinford, 1921-26.

Uni. Tromans, Stourbridge Road, Lye, 1921. Stamber Mill, Lye, 1930-35.

Unicorn Inn, Bridgnorth Road, Wollaston. A home-brew house established in 1859 by Joseph Lakin. Acquired by James Billingham in 1912 and passed on to his son, Horace James, in 1929. It was sold to distant relatives, Batham's, in 1992.

James Walters, Park Street, Lye, 1930-39.

Adam Wassell, Heath Lane, Old Swinford, 1921-35.

Frederick Webster, 78 Brettell Lane, Amblecote, 1926.

Henry White, Pedmore Road, Lye, 1914-20.

Isaac Whitehouse, Stourbridge Road, Lye, 1914.

Robert Woodhouse, Hay Green, 1872.

George Wooldridge, Park Street, Lye, 1914. Mrs Sarah Wooldridge, Park Street, Lye, 1921.

Mrs Mary Wooldridge, 14 New Road, 1914-21. Walter Wooldridge, 14 New Road, 1923-30.

Timothy Worrall, Stourbridge Road, Lye, 1914. 80 Lower High Street, 1930-39. Mrs Esther Worrall, Lower High Street, Amblecote, 1940.

Jabez Wylde, Upper High Street, Lye, 1914-20.

TIPTON

Albion Inn, Horseley Heath. 'Fine Home Brewer Ales' was painted across the front of this house, owned by W. Simpson during the first decade of the twentieth century. Simpson also sold Whitbread's London Stout and India Pale Ale. He also owned a retail brewery at Eve Hill, Dudley.

Mrs Alice Allsop, 62 Park Lane, West, 1914, and Groveland Road, Dudley Port, 1923.

Thomas Allsop, Groveland Road, Dudley Port, 1914.

Prince's End Brewery, Tipton.

Bloomfield Home Brewery.

George Ashford, 194 Bloomfield Road, 1914.

John Ashmore, 10 Canal Street, 1914.

Arthur Bagnall, 28 Ocker Hill Road, 1914–23.

Beehive Brewery, Lower Green. A.& J. Sherwood in 1939.

Mrs Alice Bloomer, 74 Park Lane West, 1914.

John Brown, Sedgley Road West, 1914.

California, George Street. A home-brew house originating in 1850, when William Hampton was its landlord. It was later owned by the Lloyd family. It and their other pub, the Four Ways in Brown Street, were taken over by W&DB in 1941. The California closed in 1973.

George Challinor, 186 Dudley Port, 1914–23.

Isaac Chater, 54 Wood Street, 1914–21 and Park Lane West, 1923.

Albert Clarke, 176 Horseley Heath, 1914.

William Henry Cox, High Street, 1914–21.

Theodore Crawford, Bloomfield House Brewery, Bloomfield Inn, 54 Bloomfield Road/ Dudley Road, 1907. Bottled beers sold as Knights Ales.

Ernest Day, 12 Dudley Port, 1921.

Samuel Edwards, 6 Lock Side, 1914–20.

Arthur A. Fellows, 194 Bloomfield Road, 1921–23.

Robert Henry Fitzsimons, Canal Street, 1914–23.

Fountain Inn, 3 Dixon's Green. A home-brew house established by 1830 with Thomas Paskin as landlord. It was sold to M&B by the Evans family in the 1920s and closed around 1974.

Four Ways Inn, 27 Brown Street, Kate's Hill. A home-brew house, established by 1870 under the Lloyd family. The brewery and pub were taken over in 1941 by W&DB, whereafter brewing stopped.

Fox, Gospel Oak. A home-brew house by 1845, when Isaac Caddick was landlord.

Benjamin Garbutt, 4 High Street, Prince's End, 1914–35.

William George, 45–6 Burnt Tree, 1921–40.

Mrs Phoebe Gould, 49 Alexandra Road, 1914. Sidney Gould, 49 Alexandra Road, 1921–40.

George Griffiths, 33 High Street, 1914-21.

Hearty Good Fellow, 9 The Square, Woodside. A home-brew house built prior to 1860, when it was owned by the Pearson family. Who retained it up to 1900. Charles Simmonds brewed here during the 1920s. This house closed in 1938.

William John Hill, 81 Bloomfield Road, 1914-23. Mrs Eliza Ann Hill, 81 Bloomfield Road, 1930-35.

Mrs Elizabeth Hunstone, Upper Church Street, 1923-35.

Mrs Susan Hunt, 56 Owen Street, 1921.

Charles Jackson, 10-11 Waterloo Street East, 1914-23.

Samuel G. Jones, Sedgley Road West, 1914-20.

King William, 9 Cole Street, Darby End. A home-brew pub established pre-1870. Daniel Batham, founder of Batham's Brewery, was landlord here at the opening of the twentieth century. He sold the building to Hanson's in 1915, but retained the license until 1921. The King William was rebuilt in 1956.

Leopard Inn, 25 High Street, Kate's Hill. In existence by 1870, this home-brew house was then owned by Lewis Crump. It was bought up by J.F.C. Jackson of the Diamond Brewery in 1926, when brewing there ceased. The Leopard closed in 1937.

Loving Lamb, 2 High Street, Kate's Hill. Established by 1840 under Richard Harper, this home-brew house was taken over by Darby's Brewery in 1937. It closed in 1949.

Harry Lyndon, 18 Waterloo Street, 1914-20.

Alfred Mander, 13a Walton Street, 1914-40.

William Martin, 57 Dudley Road, 1900.

Mrs Elizabeth Middleton, 37 Bloomfield Road, 1914-23.

James Mills, 26 Hurst Street, 1914-23.

Henry Morris, Conygree Road, 1914. Mrs Ann Morris, Conygree Road, 1921-30.

Old Woolpack Inn, Castle Street, Market Place. A home-brew house established in 1622. In 1881, then owner Isaac Aulton was advertising his 'Prime Home Brewed Ales'. The house was later taken over by Hanson's and brewing ceased.

Harry Onions, 118 Horseley Heath, 1914-20.

Ezra Padley, Toll End Road, 1880.

Thomas Henry Palmer, 40 High Street, Prince's End, 1914-20.

Jabez Pessoll, Upper Church Lane, 1914.

Frederick Plant, Upper Church Lane, 1914-21.

John Henry Purnell, 1 Lower Green, 1914. Alfred Purnell, 1 Lower Green, 1921-30.

Railway Tavern, The Croft, Woodside. A home-brew house formerly known as the Three Furnaces. Established pre-1870 when it was owned by Charles Hillas. It was acquired by Hanson's in 1934.

Walter Randall, 30 Aston Street, Toll End, 1914-23.

Francis Rhodes, 54 Bloomfield Road, 1914-23.

William Rich, 51 High Street, Prince's End, 1914-35.

Royal Oak Brewery, Harts Hill. Established behind the Harts Hill Tavern in about 1869 by William Smithyman.

Thomas Henry Scriven, 74 Park Lane West, 1921-35.

Alfred Sherwood, Lower Green, 1914-39.

Shoulder of Mutton, 29 Dixon's Green Road. A home-brew house established by Thomas Gwinnutt, *c.* 1850. It was taken over by Butler's of Wolverhampton in 1939 and closed down in the 1970s.

Mrs Florence Sirrell, Brown Lion Street, Bloomfield, 1921-35.

Thomas Sleeth, 12 Dudley Port, 1914.

Mrs Sarah Standford, 29 Victoria Street, Prince's End, 1914-21. Herbert Standford, 29 Victoria Street, 1923-30.

Star Brewery, 176 Horseley Heath. Albert H. Clarke, 1920.

Josiah James Stevenson, 27 Horseley Heath, 1914-21. 29, Victoria Street, 1930.

Charles Summers, 116 High Street, Prince's End, 1914. Mrs J. Summers, 116 High Street, Prince's End, 1921-39. Mrs Emma Summers, 1940.

William Summers, Brown Lion Street, Broomfield, 1914.

Thomas Taylor, 36 Union Street, 1914-20.

Three Horse Shoes, Brick Kiln Street. A home-brew house owned by Johnson & Phipps of Wolverhampton. It closed in 1958.

Albert E. Turner, Sedgley Road East, 1921.

Victoria Brewery, 23 Dudley Road, Dudley Wood. Situated behind the Victoria Inn and established before 1870, when Davis Weston was its landlord. In 1900 Harry Bridgewater was landlord. The pub was leased to Butler's of Wolverhampton in 1951. Later acquired by W&DB, it is now a free house.

Mrs Ellen Walford, 7 Park Lane West, 1914-21.

John Thomas Walker, 38 Coppice Street, 1914-21. 10 Canal Street, 1923-30. Mrs Elizabeth Walker, 10 Canal Street, 1935-40.

Henry Ward, 56 Owen Street, 1914.

George & Elisha Whitehouse, Park Lane West, 1878.

Whitehouse Brothers, Park Lane West. A merger of the above brothers in 1863. George and Elisha Whitehouse 1864-68, Elisha Whitehouse in 1880. The company eventually became John Whitehouse by 1902.

Joseph Whitehouse, High Street, 1862-68.

Charles Williams, 75 Union Street, 1914-23.

UPPER GORNAL

Alfred Ernest Allen, 1923-26.

William Henry Anderson, Sheepcote Brewery, 1888.

Edward Bodenham, 1914-21.

John Bodenham, 1923-35.

Britannia, Kent Street. Popularly known as 'Old Sal's' after landlady Sally Williams. A home-brew house, it was built in 1780. The pub had previously been run by Mrs William's parents, Thomas Peacock and his wife Sarah. The Britannia brewed up until 1959. It resumed brewing in May 1995 under Phil Bellfield, using a new three-barrel plant. Brewing ceased once more when the pub was taken over by Batham's in 1997. Some beer was reputedly brewed there in 1998.

Mrs E. Ada Cartwright, Kent Street, 1914-23.

Arthur Gibherd Cartwright, 1914-21.

Mrs Mary E. Cole, 1930-35.

William Richard Easthope, 1939-40.

Thomas Fellows, 1921.

Henry Fullwood, Sheepcot Wall, 1874.

Good Intent, Vale Street. A home-brew house originating in the mid-nineteenth century. The first known owner was Benjamin Nichols. The family remained brewing here for nearly sixty years. The house was acquired by Hanson's in 1937.

James Guest, retail brewer, 1833. James Guest & Sons, Orchard Brewery, 1872-80. James C. Guest, 1930-35.

Mary Guest, 1833. Retail brewer.

Harry Hammond, 1914-35.

Isaac Hughes, 1833. Retail brewer.

Jolly Crispin, Clarence Street. An early nineteenth-century home-brew house whose earliest landlord was John Lewis. His son William took over, and after him the Meanley family. John Foley of Kate's Hill Brewery, Dudley, bought the house in 1879. J.P. Simpkiss & Son took over in 1940 and retained it for forty-five years before they themselves were taken over by Greenall Whitley.

John Meredith, Ruiton, 1914.

Bert Middleton, 1930.

Isaac John Mills, 1914-23.

James Morris, Ruiton, 1914.

Samuel Naylor, 1833. Retail brewer.

Old Mill, Windmill Street. A home-brew pub built in 1875 by John Hyde. It was later bought up by Holden's in 1946.

David Round, 1833. Retail brewer.

John Turner, 1914.

Mrs Rachael Westwood, 1914.

William Henry Westwood, 1914-21.

WALL HEATH

Alfred Alt. Kinsey, 1921.

William A. Kinsey, Albion Street, 1923-30.

John Solari, High Street, 1914-35.

WALSALL

Albion Brewery, 1 & 2 Pool Street, 1888.

Sarah Ann Allsop, Pleck Brewery, 30 Oxford Street, 1914-20. A three-storey brewery built around 1890. Formerly situated behind the Royal Oak. The firm later became S. Allsop & Sons Ltd, with stores at Walsall Railway Station. The company was acquired by Ansell's. The Royal Oak was demolished and a new pub built to replace it. The brewery has since been demolished.

Mrs Paulina Annakin, 29 Hill Street, 1914. William Annakin, 29 Hill Street, 1921-26.

Benjamin Baggott, 31 Dudley Street, 1914-26.

Henry Bateman, Walsall Wood, 1921.

Richard J. Bayley, 126 Stafford Street, 1921-23.

Isaac Beebee, 38 Day Street, 1914-21.

Mrs Emma Bird, Butts Inn, 46 Butts Street. Originating in 1855. Mrs Bird took over

Left: *Major George Cox's Pleck Brewery, Walsall.*

Opposite, left: *The Littleton Arms, Walsall.*

Opposite, right: *The Littleton Arms Brewery.*

the business on the death of her husband. The company acquired two further tied houses. It registered as a company in 1920, but went into voluntary liquidation in 1929. See: N.F. Bird in the main listing.

Black Swan, Blue Lane. A home-brew house whose landlord and brewer in 1881 was Charles William Pearson.

Joshua Bloomer, 44 Lower Hall Lane, 1914.

Thomas Boulter & Son, Shire Oak Brewery, 1900-30.

Mrs John Boyce, 11 & 12 Windmill Street, 1914.

Joseph Armitage Brook, Butt Road, 1914-20.

I. Brookes, 2 Regent Street, Birchills, 1904.

Josiah James Brookes, Leamore, 1880.

J.W. Brookes, 17 Lysways Street, 1920.

Alfred Brown, 15 Caldmore Green, 1914-26.

Mrs Emma Brown, 58 New Street, 1914-26.

Robert Brown, 22 Tantarra Street, 1914.

Mrs Rose Bull, 106 Stafford Street, 1923-40.

Josiah Burgess, 4 High Street, 1914. Walter Burgess, 4 High Street, 1921-26.

Isaac Butler, Ettingshall Road,1861. Brewer and grocer.

Mrs Sarah Ann Butler, 30 Newhall Street, 1914-23.

Butts Brewery, The Butts, 1923.

Edward Challinor, 17 Bank Street, 1914-20. 26 Pleck Road, 1921.

William Challinor, 87 Wisemore, 1914-20.

James Cleobury, 143 Portland Street, 1914-20.

Alfred Cooke, 126 Stafford Street, 1914.

David Cooper, 9 Green Lane, Birchills,

Samuel Cooper, 90 Hatherton Street, 1864.

Sidney Cousins, 38 George Street, 1914.

Major George Cox, Walsall Brewery, Pleck, 1884.

Thomas Cox, 90 Ablewell Street, 1914-21.

John & William Cresswell, Pelsall, 1864-92.

Norman Dawson, 90 Ablewell Street, 1923-39.

Stephen Dawson, 1 Bott Lane, 1914-20.

Charles W. Done, 3 Ryecroft Street, 1914-21.

Thomas Edge, 14 Portland Street, 1914.

Mrs Mary Ann Fox, Wolverhampton Road, 1914.

Mrs Harry Green, 19 Bridgman Street, 1914. Mrs Hannah Green, 19 Bridgman Street, 1921-35.

Robert Harris, Shelfield, Walsall, 1932.

Daniel Herbert & Edward Harrison, Laburnams Brewery, Rushall, 1887-1900.

Frederick Harrison, Rushall Brewery, Rushall, 1880-1904. Daniel Herbert and Edward Harrison, Laburnams Brewery, 1904.

William Edward Hatton, Pelsall, 1900-04.

Thomas Hawley, 20 Pool Street, 1914. Mrs Kate Hawley, 20 Pool Street, 1921-23.

Horace Holmes, 193 Stafford Street, 1920. Samuel Holmes, 1921-30.

Samuel E. Holmes, Bath Street, 1921-35.

Sydney John Holmes, 38 George Street, 1921-26.

Henry Kirby, 26 Pleck Street, 1914.

William Lane, 84 Bloxwich Road, 1914-21.

William O. Letts, 101 Lichfield Street, 1914-30.

Littleton Arms, Littleton Street East. A home-brew house built post-1862. Samuel Tippett was its first licensee. William Utting was landlord from 1880-1900. There is a small two-storey brewery to the side of the pub, with louvre-windows along the top storey for cooling.

Thomas Burdett Lowe, 7 Windmill Street, 1914-20. 23 Lower Walhouse Street, 1923-40.

Mrs Ethel Marley, 11 & 12 Windmill Street, 1921.

Arthur Marshall, 11 & 12 Windmill Street, 1923-30.

Mrs Lillian Marshall, 337 Green Lane, 1914-21

Noah Mills, 44 Lower Hall Lane, 1921-23. Mrs Ada Mills, 44 Lower Hall Lane, 1930.

Charles Naylor, 8 Orlando Street, 1921.

Arthur Noake, 16 Hill Street, 1921.

David Noake, 30 Green Lane, 1914-21.

Joseph Pagett, 352 Pleck Road, 1914-20.

William Alfred Payne, 22 Tantarra Street, 1921-23.

Charles William Pearson, Black Swan, Blue Lane, 1881.

Bernard Perry, 92 Wolverhampton Street, 1914-21.

George Place, Blue Lane West, 1914-20.

Thomas Powell, 33 Lower Hall Lane, 1914-21.

James Prentis, Bradford Street, 1839.

Joseph Benjamin Preston, 92 Wolverhampton Street, 1923.

Thomas Roberts, 269 Green Lane, 1914-26.

William Roberts, Brownhills, 1900-24. Brewer of 'Entire'. William Roberts Brewery.

John Rogers, Bath Street, 1914.

William Rollins, 16 High Street, 1914.

Arthur Sheryn, 16 Hill Street, 1914.

George Sidwell, 37 Bridgeman Street, 1914-23.

Abraham Smith, 2 Regent Street, Pleck, 1914-20.

Mrs Emma Smith, 34 Queen's Street, 1914-20.

Mrs Lucy Smith, 19 Stafford Street, 1914-23. Leonard Smith, 19 Stafford Street, 1930-35.

Mrs Ann Thomas, 109 Wolverhampton Street, 1921.

George Thomas, Green Lane, Birchills, 1914-23. Frederick W.A. Thomas, 2 Green Lane, 1930-35.

John Thomas, Wolverhampton Street, 1914.

Edward Toon, 81 Lord Street, 1914-21.

John Toon, 52 Dudley Street, 1914-23.

Edward Towe, 36 Duncalfe Street, 1921.

Turk's Head Inn, Digbeth. A known home-brew house established before 1826.

G.S. Twist, 92 Wolverhampton Street, 1930-39.

Victoria, Lower Rushall Street. Established pre-1868. Samuel Nock was its first landlord and brewer. John Hill was brewer from 1914-20. There is a small tower brewery to the rear, now used as accommodation.

David Walker, 106 Stafford Street, 1914-21.

William Walkerdine, 36 Duncalfe Street, 1923.

Watering Trough Inn, 90 Ablewell Street. Norman S. Dawson, 1940.

J.Whitehouse & Son, 1920.

Miss Doris Wilcox, 173 Blue Lane West, 1930-35.

Ernest William Wilkes, 8 Orlando Street, 1914.

Matthew Wilkes, 36 Duncalfe Street, 1914.

Mrs Frances Gertrude Wilson, 173 Blue Lane East, 1914-23.

Joseph Wilson, Retail brewer, died 1869. He left his premises, stock in trade and brewing plant to his wife Mary (Walsall Local History Centre: ref. 48/12/13).

Windmill Brewery, Bath Street. Samuel E. Holmes, 1937.

Albert E. Wood, 10 Paddock Lane, 1914-23.

Arthur Yates, 39 Bank Street, 1914. Annie Yates, 39 Bank Street, 1921-23.

WEDNESBURY

Alfred Blakemore, Darlaston Road, King's Hill, 1921.

Richard Henry Butler, Piercey Street, Newtown, 1914. Mrs Arabella Butler, 1921.

Above: *An advertisement for Thomas Cartwright's Old Park Inn.*

Left: *The Victoria Brewery, Walsall.*

Thomas Butler, 89 Mill Street, King's Hill, 1914-23.

Thomas Cartwright, Old Park Hotel, Darlaston Road, 1868-80.

James Cleobury jr, Wood Green, 1921-23.

Henry Williams Clifton, King's Hill, 1914-20.

Arthur Darby, Piercey Street, Newtown, 1930.

John Dayman, Elwell Street, 1914-20.

John Dewson, Dudley Street, 1828.

F.J. Dicken, 23 Wood Green, 1888. F. Dicken jr, in 1891. Mrs Sarah Dicken, Wood Green, 1914.

John Disturnal, High Street, 1850. Corn, flour and provision dealer and retail brewer.

John Taylor Duce, & Sons Ltd, offices at Spring Head; brewery in Church Street. Wine and spirit merchants and brewers, declared bankrupt in October 1888 with debts of £1,703 11s 11d.

Arthur Fieldhouse, 81 Wood Green Road, 1930.

Benjamin Griffiths, Trouse Lane, 1828.

Harry Hodgkiss, Oxford Street, 1923.

Albert Hughes, Darlaston Road, King's Hill, 1921. Mrs Louisa Hughes, 139-140 Darlaston Road, King's Hill, 1923.

John T. Jackson, 37 High Bullen, 1904.

Thomas Jones, 37 High Bullen, 1868-84. William Jones, 1888, H. Jones, 1900.

W.E. Jones, High Bullen, 1884.

John Lacey, 11 Bilston Road, 1914-21.

Nicholas Edmund Lacey, Portway Road, 1914-20.

Charles Henry Lloyd, Piercey Street, Newtown, 1923.

George Lloyd, 3 High Street, 1884.

C.A. Loxton, High Street, 1839.

Joseph Maloney, Woods Bank, Moxley, 1914-21.

Samuel Middleton, Foundry Street, King's Hill, 1868.

Millward Brothers, Lea Brook, 1910.

Thomas William Nicholls, Trouse Lane, 1914-20.

Mrs Sarah Onions, 14 Camp Street, 1914-40
George Peters, Dudley Street, 1921.
Edward Phillips, Dale Street, 1906.
Isaiah Platts, 74 Darlaston Road, King's Hill, 1921.
William Poxon, 100 Holyhead Road, 1921.
Prince of Wales, King's Hill. Home-brew house originating with ex-gunmaker, Samuel
   Streatham. His son, Joseph inherited the pub following his death in 1884. The pub was
   later bought by Lashford's and in 1971 by Courage. Leased to Holden's in 1980.
John Ricketts, Holyhead Road, 1914. Mrs Mary Ricketts, 100 Holyhead Road, 1923.
Mrs Emily Rogers, King's Hill, 1914-20.
Thomas Stone, Trouse Lane, 1828.
George Till, New Street, 1914-20.
Samuel Tonks, 28 Elwell Street, 1912-21.
Ann Turner, Wood Green, 1828.
Enoch Turner, King's Hill Fields, 1828.
John Wainwright, 3 High Street, 1888.
Mrs Sarah Wheale, Queen Street, 1914. Fred Wheale, Queen Street, Moxley, 1921.
William Whitehall, 106 Franchise Street, 1914-35. Robert Whitehall, Franchise Street,
   1939-40.
Joseph Whitehouse, Spring Vale House, Russell Street, 1872-84.
Mrs Mary Louisa Whittaker, 56 Russell Street, 1914-20.
Harry Howard Wilkes, Old Park Road, 1914-23.
Edward Williams, 47 Foster Street, 1914.
William Winsper, 47 Foster Street, 1921.
Alfred Woodcock, Elwell Street, Newtown, 1923-40.
Ernest Woodhall, Holyhead Road, 1914-40.
Peter Woodhall, King's Hill Fields, 1828.

WEDNESFIELD

Joseph Beech, Wood End, 1930-40.
John Gregory, High Street, 1914-23. Ernest John Gregory, 1930-40.
Robert Thomas Griffiths, Lichfield Road, 1914-23.
Wallace Johnson Gumbley, Church Street, 1914-23.
Thomas Howe, High Street, 1921-30. James Howe, High Street, 1930-39.
Walter Lane, retail brewer and key maker, 1833.
Luke Nicholls, retail brewer and rat-trap maker, 1833.
Richard Tomkys, Wednesfield Heath, 1833.
William Warner, High Street, 1878-84.
Mrs Hannah Williams, High Street, 1914. Richard T. Williams, High Street, 1921-30.

WEST BROMWICH

Frank Archer, Hall Green, 1914-21.
G. Arnold & Co., Dartmouth Park Brewery, New Street, 1908-23.

BATES' SPONWELL ALES.

"STRATHAVON"

A Special Blend of Old Scotch Whisky

Sponwell Brewery, West Bromwich.

ESTABLISHED 1865.

Bates' Sponwell Brewery, West Bromwich.

Charles Bailey, 27 Dartmouth Street, 1908-23.

Thomas Henry Bates, 119-121 Spon Lane, 1900-12. Henry Bates, 1914-23.

Thomas Beddard, Albert Street, 1914.

Boat Inn Brewery, Gold's Hill, Joseph Jones, 1920.

William Bowen, 2 Oldbury Road, Greets Green, 1914-35. William Bowen Ltd, 1939-40.

Mrs Emma Brain, Great Bridge Street, 1914-20.

Edwin Capewell, Albert Street, 1921.

John Amos Cowles, retail brewer of 34 Bratt Street. In August 1887 he was fined £25 for attempting to defraud the Inland Revenue by adding sugar to increase its specific gravity after he had paid duty at a lower rate.

Joseph Cox, Albert Street, 1923-26.

William Cox, retail brewer of 140 Lodge Road. He was fined £25 for attempting to defraud the Inland Revenue by adding sugar to his brew.

Crown & Anchor, a home-brew pub. In October 1881, its then brewer landlord, George Gough, was declared insolvent.

Enoch Dabbs, Harvell1s Hawthorn, 1914.

Frank William Davis, Gold's Green, 1921-23.

James Downing, Albion Road, 1914-39 and at 73 John Street in 1923.

Thomas Downing, 73 John Street, 1914-23.

Thomas Edwards, 29 All Saints Street, 1914-26.

George Faulkner, 45 Lyndon Street, 1914. John Faulkner, 45 Lyndon Street, 1920.

John Griffiths, New Street. Retail brewer, 1850.

Mrs Sarah Hale, 79 Moor Street, 1914-20.

Mrs Annie Hampson, Holloway Bank, 1923-35.

Dennis Harris, 62 Harvell's Hawthorn, 1923.

Edwin Holden, Swan Village, 1920.

William Hollyhead, Union Street, 1920.

ARTHUR J. PRICE,

LEWISHAM BREWERY,

WEST BROMWICH.

PALE AND MILD ALES AND STOUT.

Price Lists on application.   Telephone: No. 104 W.B.

A.J. Price's Lewisham Brewery, High Street, West Bromwich.

Susannah Hyde, Swan Village, 1920.
Mrs Annie Jinks, Holloway Bank, 1914.
Harry Jones, Hill Top, 1914-21.
Joseph Jones, Gold's Green, 1914.
Thomas Moorhouse, Hargate Lane, 1921.
Joseph Arthur Parker, 58 Bull Street, 1914-35.
George Parkes, Stoney Lane, 1914-30.
Frederick Ernest Perks, 62 Harvills Hawthorn, 1926.
Arthur James Price, 45 High Street, 1900-08.
James Price, Oak Road, 1839.
Walter Randell, Gold's Green, 1926.
Spencer's Phoenix Brewery, 335 High Street, 1908-28.
James Stanton, Braybrook Street, 1914-35.
Mrs Florence Stockley, Hill Top, 1921.
Harry Thomas, 62 Harvell's Hawthorn, 1821.
William Twist, 150 Union Street, 1914-20.
Union Cross Inn, Greet's Green. Daniel Williams, ale and porter brewer, 1850.
George & Elisha Whitehouse, Church Lane and Park Lane West, Tipton.
Joseph Whitehouse, 23-25 Ryder Street, 1914-20.
Thomas Wright, 63 Horton Street, 1914-20.

WILLENHALL

Thomas Allen, 86 Walsall Street, 1914-26.
Barnaby Bailey, New Invention. Retail brewer, 1833.
William Booker, The Crescent, 1914-21.
Charles Collett, 29 Union Street, 1878-80.
George Deans, Lane Head, Short Heath, 1914-21.

Archibald Fenn, Lane Head, Short Heath, 1914-21.
Humphrey Fox, John Street, 1833.
George Henry Harbach, 118 St Ann's Road, 1914-20.
Maria Hartill, Chapel Green, 1833.
Randle Hobley, 57 Walsall Street, 1914-30.
William Lees, John Street, 1833.
Sefus Lowbridge, 93 Coltham Road, 1914-23.
Henry Mills, 150 Walsall Road, 1921-30. See: West Midland Brewery.
Frederick William Minors, Cheapside, 1914. Cross Street, 1920.
Richard Pedley, Wolverhampton Street, 1833. Also a shoemaker.
Benjamin Smith, Wolverhampton Street, 1833. Also a locksmith.
Frederick Turberville, Lane Head, Short Heath, 1914-23.
Geoffrey Wakelam, Temple Barr, 1914.
Sampson Wakelam, 150 Walsall Road, 1914.
Mrs Elizabeth Watkins, 2 Wednesfield Road, 1914-20.
Samuel Whitehouse, 2 Water Glade, 1914-21.
Richard Wilkes, Short Heath, 1833. Also a locksmith.

WOLVERHAMPTON

Joseph Adams, Skidmore Road, Bradley, 1930.
Lucy Adshead, Bilston Street, 1833. Retail brewer.
John Allman, Fourth Lock House, Birmingham Canal, 1833. Apparently serving the
   canal boatmen.
David Bate, Stafford Street, 1839.
Elijah Bagot, Horseley Fields, 1833. Retail brewer.
Benjamin Baker, Bell Street, 1833.
Sarah Baker, retail brewer, 1833.
John Barlow & Co., 21 Upper Villiers Street, 1900.
George Bate, Bloomsbury, 1833.
Thomas Bishton, Great Brickiln Street, 1833.
Brewery Tap, Dudley Road. Holts Entire was brewed here briefly.
Cyril T. Brewster, 15 Lewis Street, 1923-26.
Sarah Brierley, Salop Street, 1833.
George Brooks, Canal Street, 1839.
William Howard Broome, 194 Wolverhampton Road, Heath Town, 1923-26.
Richard Brown, Grove Street, Heath Town, 1902.
Thomas Burey, Horseley Fields, 1833.
Charles Burton, Duke Street, 1833.
Edward Butler, Pountney Street, 1833.
Abraham Cartwright, Tanhouse Lane, 1833.
Catherine Cartwright, Tettenhall Wood, 1914-20.
William Chamberlain, Berry Street, 1833.
William Childe, Wheelers Fold, 1839.
George William Chilton, 14 St Mark's Street, 1921.
Frederick Coleclough, Coven, 1904-12.

Francis Coles, New Street, 1833.

Louis Connolly, Mary Ann Street (Brewery), Chapel Ash and 42 Worcester Street, 1914.

Mary Cotterell, Stafford Street, 1833.

Thomas Cotton, Wheelers Fold, 1833.

Christopher Craddock, Bond Street, 1833.

George William Crane, 167 Horseley Fields, 1914-20.

William Cropp, Charles Street, 1833.

John Ignatius Cunningham, 67-68 Moore Street, 1908.

Aubrey Cupiss, 325 All Saints Road, 1914-21.

Bernard Dagnan, 17 Inkerman Street, 1923.

Samuel Davies, Stafford Street, 1833-39.

Edward Dawes, Cornhill, 1839.

John Dawes, Newbridge Brewery, Tettenhall Road, 1880.

Thomas Dean, Stafford Street, 1833.

Joseph Edwards, Salop Street, 1833.

J. Edwards, Great Brickiln Street, 1833.

William Edwards, Dudley Road, 1833.

George Evans, Warwick Street, 1833.

Edward Fellowes, Elliotts Row, 1833.

George Fieldhouse, Stafford Street, 1833.

Thomas Fieldhouse, Wright Street, Bradley, 1920.

Isaiah Fisher, 31 New Hampton Road, 1884.

John Foster, 149 Coleman Street, Whitmore Reans, 1880.

Frantz & Perry, High Street, 1874.

Henry Freeth, 58 Wulfruna Street, 1914-21.

John Gallagher, St John's Square, 1851.

William Henry Garnett, 227 Coleman Street, 1914-20.

Mrs Ann Gaukrodger, Tettenhall Road, 1914.

The George, St James' Square, 1833. William Walker, landlord and brewer.

Arthur Edward George, Salop Street, Bradley, 1921.

Henry George, Cross Street, Bradley, 1921.

Charles Giles, Essington, 1878-88.

Francis Gill, 19 Piper's Row, 1923.

Edward Glover, Horseley Fields, 1839-45.

J. Glover & Sons, King Street and at Sutherland Road, Longton, 1864.

William Goodman, Staford Street, 1839.

Miss Mary Jane Gough, 102 Low Stafford Street, 1914.

Grapes Brewery, Chapel Ash. Louis Connolly, 1914.

Edward Greene & Son, Exchange Vaults, Red Lion Street, 1884. Greene, King & Sons Ltd.

Frederick York, manager, 1900-10.

John Griffiths, Coach & Horses Brewery, Bilston Road, 1884. Griffith & Cattell, 1888-92.

Joseph Griffiths, Salop Street, Bradley, 1921.

Thomas Groves, 14 St Mark's Street, 1914.

Henry John Haddock, 194 Wolverhampton Road, Heath Town, 1914-21. 58 Wulfrun Street, 1923-26.

Ernest Harper, 19 Brickiln Street, 1914.

Richard Harper, Salop Street, 1833.

William John Hawkes, 7 Cross Street, Bradley, 1940.

William Haynes, Horse Fair, 1833.

George Hill, 182 Park Street South, 1912-14.

Thomas Hill, North Street, 1833.

Thomas Hill, Warwick Street, 1833.

Edward Holland, Dudley Street, 1833.

Thomas Holland, Goldthorn Hill, 1833.

J. Hood, Oxford Street, 1833.

Benjamin Hook, Horseley Fields, 1833.

Joseph Horabin, North Street, 1833.

William Horton, Union Street, 1833.

James Howell, King Street, 1833.

Joseph Howell, King Street, 1833.

Charles Thomas Hughes, 19 Piper's Row, 1921.

Henry Hughes, Worcester Street, 1833.

Harry Instone, 15 Bishop Street, 1914-23.

James James, Stafford Street, 1839.

Betsy Jeavons, Bank Street, Bradley, 1920.

Albert Henry Johnson, 1 Moore Street South, 1914-20.

Frederick William Jones, 19 Piper's Row, 1914.

George Jones, Tettenhall Road, 1921.

Henry Kent, 28 Great Hampton Street, 1923.

Charles Howard King, St James' Square, 1851. Bernard Place, Darlington Street, 1874.

Lamb Inn, Horse Fair. William Webb, brewer, 1839.

Lamsdale & Eccleston Ltd, Merindale Street, 1908.

Edward Lawrence, & Co., Bilston Road, 1908. Lawrence was declared bankrupt in 1910.

Alfred & Edward Leary, 6 Horseley Fields, 1900.

Mrs Charlotte Lewis, 32 Eagle Street, 1914-23.

John Lockley, 40 Ash Street, 1914-21. Anne E. Lockley, 40 Ash Street, 1923-35.

Joseph Lodge, Bond Street, 1851.

John McDonald, 21 Stafford Street, 1921-23.

Joseph Martin, 21 Franchise Street, 1914-39, and at Ash Street in 1935.

Fred H. Mason, Springfield, 1926.

Benjamin Meanley, Graisley Hill, 1833.

Lloyd Mears & Co., Home Ale Brewery, Ettingshall, 1904-14.

Samuel Meek, Newbridge, Tettenhall, 1861-64.

Merridale Brewery Co., 132 Merridale Street, 1914.

Midland Brewery, Bilston Road. Harmer & Co. (1912) Ltd. Acquired by Ansell's.

Edward Millington, Willenhall Road, 1833.

William Morris, Duke Street, 1833.

William Moseley, 132 Merridale Street, 1914-40.

P.Y. Mottram & Co., Market Street, 1851.

John Nash, Berry Street, 1878.

James Nicholls, Piper's Row, 1833

The Queen's Arms, Graisley.

Richard Nicholls, retail brewer, 1833.
Daniel Paul, 74 Hill Street, Bradley, 1930.
Alfred Richard Perry, 50 Salop Street, 1914-30.
Thomas Perry, Stafford Street, 1833.
Thomas Perry, Little Berry Street, 1833.
Phillips Brothers, Clarence Street, 1874. N.T. Lloyd, agent.
Thomas Picken, 20 Wharf Street, 1914-20.
William Poole, 72 Canal Street, 1880.
John E. Preston, 102 Lower Stafford Street, 1921.
Edward Pritchard, Green Lane, Dudley Road, 1833.
George Pugh, Horseley Fields, 1839.
Queen's Arms, Graisley Row. See: Shelton Brewery in main list.
Benjamin Ravenscroft, Goldthorn Hill, 1833.
Reade Brothers & Co. Ltd, Cleveland Road, 1900.
James Richards, Can Lane, 1833.
Joseph Richards, Great Brickiln Lane, 1833.
John Rogers, 70 Steelhouse Lane, 1914. John Rogers & Calcutt Ltd, 70 Steelhouse Lane, 1921-24. Brewery at Eagle Street from 1923.
Mrs Elizabeth Rollings, 10 Cartwright Street, 1921-23.
William Rudge, Salop Street, 1833.
Thomas Salt, Salop Street, 1833.
Thomas Salt & Co., 15 Snow Hill, 1874. John Sills, agent. Thomas Salt & Co. Ltd, 1914.
John Salter, Walsall Street, 1833.
Henry Wilson Sambrook, 28 Great Hampton Street, 1914-23.

Ben Shaw, 10 Cartwright Street, 1914.

Samuel Sherwood, Brickiln Lane, 1833.

Edward Shore, 10 Cartwright Street, 1935.

Richard Simpkins, Waterworks Lane, Tettenhall, 1880-88.

Ephraim Slynn, Canal Street, 1833.

Charles C. Smith, 1 Little Brickiln Street, 1872-80. Also at Bishop Street post-1884 and New Zoar Street from 1888.

Joseph Smith, Middle Row, 1833.

William Smith, Dudley Street, 1833.

Richard Smithwaite, Salop Street, 1833.

John Spink, Union Street, 1833.

George Spring, Stafford Street, 1833.

William Spruce, Pountney Street, 1833.

Ann Stanley, Oxford Street, 1833.

Elijah Stott, 17 Navigation Street, 1914-20.

Stratton & Co., 51 Lichfield Street, 1904.

H. Street, Eliots Row, 1833.

Samuel Stubbs, Little Brickiln Lane, 1833.

Richard White Tamlyn, Bilston Street Bridge, 1851.

Tettenhall Wood New Brewery, Jacob Cartwright, 1920.

Arthur Thompson, Newbridge Brewery, 1878. See also: John Dawes.

Tress & Turner, Market Street, 1880.

William Walker, Middle Row, 1839.

Mary Walton, Mary Ann Street, 1833.

George Ward, Darlington Street, 1833.

Thomas Ward, Skidmore Row, 1926.

Jonathon Webb, Canal Street, 1839.

William Webb, Horse Fair, 1833.

Charles Webberley, 8 Pearson Street, 1914-20.

Reuben Whitehouse, 88 Great Brickiln Street, 1914-23.

Samuel Richard Whitehouse, 20 Pountney Street, 1914-20.

George Wiley, Summerhill Lane, 1833.

Robert Wiley, Princess Street, 1833.

John Williams, Stafford Street, 1851.

Thomas William Williams, Springfield, 1926.

John Wilson, Salop Street Gardens, 1833.

James Worrall, Walsall Street, 1833.

John Charles Wright, Charles Street, 1833-39.

William Yates, Bilston Street, 1833.

J. Yearsley, Sun Street, 1864.

T. York, Dudley Road, 1864.

F. Joe Zeller & Son, Bell Place, 1908. 434 Dudley Road, 1910.

WOODSETTON

Charles Box, Regent Street, 1923–35.

Brook Inn Brewery, Sedgley Road. Taken over by Julia Hanson in 1940, whereafter brewing ceased.

Walter Foster, 1914–20.

Edwin Holden, Swan Village, 1914. See: Holden's Brewery Ltd.

Mrs Susannah Hyde, Swan Village, 1914.

Evan Shaw, 9 Brook Street, 1914–21.

Charles Turley, 1914–26.

Emanuel Whitehouse, Regent Street, Swan Village, 1914–21.

William Willetts, Brook Inn, 1930–39.

WORDSLEY

Joseph Arthurs, High Street, 1921.

Thomas J. Banks, Brierley Hill Road, 1914. Brewery Street, 1920.

Mrs Jane Dalrymple, The Green, 1914–20.

Charles Davies, The Green, 1914.

James Eccleston, High Street, 1914.

Isaac Fletcher, Audnam, 1923.

Daniel Gill, Buckpool, 1914. Mrs Daniel Gill, 1921–23. Mrs Emma Gill, 1930.

Edward Gough, High Street, 1914. High Street, 1914.

Richard W. Griffin, High Street, 1921–23.

Samuel Hill, Buckpool, 1900–14.

William H. Morgan, The Green, 1914–20.

Mrs Elizabeth Munday, 1914.

William Nicholls, High Street, 1926.

E. Oakes, High Street, 1868.

Joseph Pargeter, 1914.

Thomas Albert Parrish, 1921–30. Albert E. Parrish, 1939–40.

Joseph Price, 1921.

Andrew Smart, The Green, 1923–40.

William A. Thompson, 1914–21.

W. Wallis, High Street, 1923.

William H. Walters, Brierley Hill Road, 1921.

Mrs Sarah Wilcox, 1914.

THREE

THE RISE
OF THE
BIG BREWERS

It seems that the Black Country felt that it did not need a commercial or common brewer back in the early nineteenth century. The Dudley Brewery, founded in 1805, struggled for most of its life. Most pubs brewed their own beer. They had their own plant, and even if they did not brew themselves, a local freelance brewer, particularly around the Dudley-Netherton area, could always be found to come in and brew for them. So, why did they need to buy someone else's beer? In most cases the rise and success of the big breweries was directly linked to the tied house system. If publicans could not be found to buy their beer, then the answer was to buy pubs in which their beer could be sold.

Walter Showell was the first Black Country brewer to see this. Showell, a Birmingham man, opened a small brewery in Simpson Street, Oldbury. Slowly but surely his business expanded and in 1870 he opened a second and larger brewery at Langley, Oldbury. In conjunction he started buying up pubs when they came onto the market. As profits came in, he invested directly back into expansion. In 1884 he opened a second ninety-quarter brewery. In 1890, to raise capital, Showell's registered as a limited liability company. By now they had sixty-three tied houses in the West Midlands area. In the Parliamentary returns of the ownership of licensed houses (Parliamentary Papers 1890-91 [C28] LXVIII) Showell's houses within Birmingham and the Black Country are numbered as follows:

Birmingham	40
Dudley	2
Kingswinford	5
Oldbury	4
Rowley Regis	6
Walsall	2
West Bromwich	4

Showell's Brewery was a founding member of the Brewers' Investment Corporation. This institution's aim was to buy-up ailing pubs, secure licenses and generally to improve the image of the local brewing industry. Walter Showell's son, Charles, was later successful in buying out the other members of the Corporation, thus acquiring a further forty public houses for Showell's, who now

WHAT IS "I.P.A."?

SHOWELL'S

Indian Pale Ale in Bottles

AND NO OTHER.

SHOWELL'S Celebrated Ales and Stout in Bottle and

in Cask to be obtained at these Bars.

Showell's Brewery Co.

LTD.

157, Great Charles Street,

Birmingham.

W. KENDRICK,
Ale, Stout and Coal Merchant,
2, OLD SQUARE, WARWICK.

Truck Loads Best Selected House Coals, with Colliery Overweight, to any Station.
Clean Wigan Nuts and Washed Breeze.

WALTER SHOWELL & SON'S CELEBRATED
CROSSWELLS ALES AND STOUT.

"The CROSSWELLS BREWERY has nothing old about it except its reputation. In its size, in the extent of its operations, in the completeness and perfection of its arrangements, it is alike without peer in the most central district of the Midlands."

"The CROSSWELLS Household Ales, as they are termed, will be found as pure as unimpeachable water, the finest malt, and the choicest light-coloured hops can produce by the most exact of processes. They have no astringency or unpleasant bitterness, but, on the contrary, possess *qualities which place them in the front rank of simple and inexpensive digestive tonics.* No far-fetched simile is needed to recommend the products of the CROSSWELLS BREWERY. Their best testimonial is their popularity; indeed, if they go on increasing in public favour, a well-known phrase will have to be altered, and will read for the future as "Familiar in men's mouths as CROSSWELLS ALES."

"*Bright in colour, delicious in taste, ... a sparkling and invigorating tonic, the like of which have never been found in the physic-chest of the World.*"—Birmingham Daily Post.

Ten per cent Discount for Cash with order or on delivery; Five per cent for Cash in a month.
REPRESENTATIVE FOR WARWICK AND DISTRICT:—
W. KENDRICK, Ale, Stout and Coal Merchant,
2, Old Square, Warwick. Ale Stores, Castle Street.

Above: *Advertisement for Showell's Brewery.*

Far left: *Showell's Brewery, Oldbury.*

Left: *Beer labels, Showell's.*

styled themselves as a Birmingham company. He had succeeded his father as Managing Director in 1887. Highly respected, he was at one time Chairman of the Brewers' Association. He was also instrumental in Arthur Chamberlain's pub license surrender scheme in Birmingham [see the companion volume, *Birmingham Breweries* by the author]. Charles Showell also helped to establish the School of Brewing at Birmingham University. By 1896 however the company was facing stern competition from other rising Birmingham and Black Country brewers who had learnt their lesson from Showell's. Companies like Holt's of Aston, Ansell's, also of Aston in Birmingham, Mitchell's of Cape Hill, Smethwick and Butler's of Springfield, Wolverhampton, as well as the newly emergent Wolverhampton & Dudley Breweries combination were set to overtake them. In a bid to raise more money, the Stockport Brewery which Showell's had earlier acquired in order to gain their tied houses was sold off for £250,000. By 1914 the company was finished; it lacked the ruthless streak of modern business. The company and its 186 tied houses and thirty-five off-licences were taken over by Samuel Allsopp & Son of Burton-on-Trent. The former head of the company, Charles Showell, died in 1915.

William Butler & Co.'s rise carefully mirrored that of Showell's, but curiously the 1890 benchmark of the Parliamentary Report shows a surprising lack of tied houses. It would appear that Butler's were successful in tying into the free-trade market. There are tales of the early days of the company, before they removed to the Springfield Brewery in Wolverhampton, about how William Butler used to deliver barrels of beer by handcart to local pubs. By 1872 the company was employing Mr W. Grattridge as a traveller and agent. A second agent, George Parkes, was later taken on. By 1879, business had trebled. In 1891 the company became registered as W. Butler & Co. Ltd, with assets of £1 million. Butler died on 10 March 1893, and elder son, William Bailey Butler, took over. The brewery changed tactics, and now began buying up pubs. They began in 1900 by taking over the licensed houses of F.W. Plant of Bilston.

Left: *W. Butler, father and son.*

Below left: *W. Butler & Co., trademark.*

Below centre: *Beer label, Butler's Stout.*

Below right: *Butler's Springfield Brewery, 2004.*

Henry Mitchell jr, later to form a partnership with William Butler of Birmingham, took over his father's home-brew house, The Crown, in Oldbury Road, Smethwick, in 1861. He produced a light mild ale which he sold not only at his own house, but to other local pubs too. Due to demand he built a large brewery next door to the Crown. The building was opened in 1866. As demand further increased he was obliged to consider building a larger brewery. He discovered a likely 14-acre site at Cape Hill, overlooking Birmingham, and though it was in Smethwick, he always gave the brewery address as Birmingham. In March 1877 an artesian well was sunk and work started on the construction of the new state-of-the-art brewery. The first brew allegedly took place in July 1879. Due to the initial capital outlay, Mitchell had taken on Herbert Bainbridge as a partner to help defray the costs. The business was incorporated as a private company in 1888. By this date the firm was employing 271 people and producing an annual output of 90,000 barrels. Their amalgamation in 1898 with the (similarly named, though not related) William Butler's Crown Brewery of Broad Street, Birmingham, made the new company, Mitchells & Butler's, an instantaneous major player in the regional brewing industry. What signalled them out was their robust business drive. Alfred Homer's Vulcan Brewery of Aston, and Evans' Brewery of Perry Barr, Birmingham, were both bought up the following year in order to gain their tied houses. Expansion increased to supply the extra tied houses, and between 1912-14, No.2 Brewery at Cape Hill was built. It was capable of producing an additional 30,000 barrels a week.

Meanwhile over in Walsall, the Highgate Brewery Co. was launched in 1898. It was nearly twenty years later before they took over their first brewery, Yardley & Ingrams of Bloxwich. Cheshire's Brewery, just down the road from M&B, registered as a company in 1896. They branched out into Birmingham with the purchase of Threlfall's six tied houses. Clearly the expanding industrial city was the market to capture; but six pubs was hardly going to set the

Above: *Mitchell's Ales.*

Right: *Cape Hill Brewery, c. 1930.*

Far right: *M&B beers, 1929.*

Above: *Horse-drawn drays at M&B, 1929.*

Right: *Highgate Brewery.*

Below right: *Cheshire's Windmill Brewery.*

Below: *Highgate beer mat.*

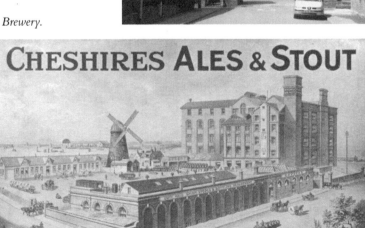

Top: *Advertisement for Cheshire's beers.*

Above left: *Wolverhampton & Dudley Breweries.*

Above right: *Banks' Mild Ale.*

Left: *A Banks' horse-drawn dray*

brewing world ablaze. As early as 1890 Mitchell's, before their merger with Butler's, had eighty-six tied houses in Birmingham. The newly created Wolverhampton & Dudley Breweries had dipped their toes into the industry, and had two tied houses. Small beer perhaps, but they were destined to become a significant player.

The company came about in 1890 with the merger of three companies, Banks & Co., C.C. Smith's Fox Brewery of Wolverhampton and George Thompson & Sons' Dudley and Victoria Breweries. Thompson was the driving force, and Banks of Wolverhampton was to become their principle brewery. Indeed the company is more popularly known locally as Banks. The building of a new sixty-quarter plant at Banks' Wolverhampton brewery signalled their entry into the West Midlands brewing world. In 1909 they took over North Worcestershire Breweries Ltd of Stourbridge and Brierley Hill, and in 1912 John Rolinson's Netherton brewery. The following year the Kidderminster Brewery with 126 tied houses was taken over. In 1917 the City Brewery Co. of Lichfield was bought out with its 200 tied houses. By 1920 W&DB were M&B's most serious Black Country rivals. By this time however, M&B and Aston brewers Ansell's had the City of Birmingham sown up [see Birmingham Breweries]. There was, it is said, an unwritten agreement that the three brewers would not 'invade' each other's territory, and while W&DB had been locked out of the city, both Ansell's and M&B actively sought new tied houses in Banks' territory.

In 1923 Butler's began a massive expansion. They bought up the brewery and licensed houses of J. Downing of Dudley, followed soon after by the purchase of the Bloxwich Brewery at Walsall. In 1925 the Cannock Brewery and its tied houses were acquired, and Eley's Stafford Brewery was taken over in 1928. Just four years before, Eley's had spent £101,989 in buying up additional properties. In the post-war period Butler's bought up the licensed premises of William Bowen Ltd of West Bromwich as well as J.A. & A. Thompson Ltd of Oldbury and A.H. Clarke of Wellington in Shropshire. James Pritchard & Son of Darlaston and Radcliff & Co. of Kidderminster were taken over. By 1953 the nominal capital of Butler's was set at £1.5 million. Their freehold and leasehold houses were valued at £3,337,556.

The last remaining major brewery, Julia Hanson & Sons of Dudley, registered as a company in October 1896 to acquire the brewery of Edward Cheshire at Birmingham. The following year they opened a new brewery in Dudley. The brewery took its name from Julia Hanson, née Mantle, whose father was landlord of the Stew Pony & Foley Arms, near Kinver. She married Thomas Hanson, a wine and spirit merchant, in 1846. They opened a wine and spirit merchant's shop in Priory Street, Dudley, the following year. A couple of years later Hanson went into partnership with maltster and pub proprietor William Hughes of Tower Street, Dudley. By 1864 Hanson was in sole charge of the business. Following his death in 1870, Julia took on the running of the company. She started buying up premises. On coming of age her two sons, Thomas and William, were put in as landlords of the Talbot in Wolverhampton Street and the Brown Lion in the Market Place. Julia Hanson died in 1894. A year later her two sons bought the old Peacock Hotel and brewery in Upper High Street, Dudley. The brewery was extensively re-built. By 1919 the brothers had over 100 tied houses and acquired more when they took over Frederick Tandy's Kidderminster Brewery that same year. They gained a further sixty tied houses, including the prestigious Stew Pony & Foley Arms, formerly run by their grandfather, John Mantle, when they took over Smith & William's Town Brewery in Round Oak in 1934. This brought their stable of tied houses up to 200. In 1936 they took over Samuel Woodhall's Borough Brewery in High Street, West Bromwich, and promptly closed it down; the object of the exercise being to gain its tied houses. In their incorporation however they had left themselves exposed. W&DB had been buying up shares in the company over the years, and by 1943 they had acquired a controlling interest.

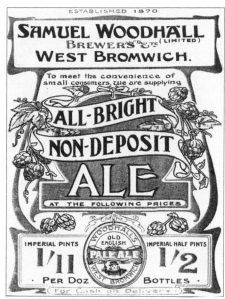

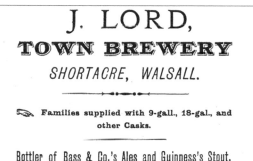

J. LORD,

TOWN BREWERY

SHORTACRE, WALSALL.

Families supplied with 9-gall., 18-gal., and other Casks.

Bottler of Bass & Co.'s Ales and Guinness's Stout.

Above left: *A Hanson's House at Gornal.*

Above: *Samuel Woodhall's Brewery.*

Left: *Lord's Brewery.*

In 1924, J.A. Fletcher's Highgate Brewery merged with John Lord's Brewery to form Walsall Breweries Proprietary Ltd. This combination took over Arthur Beebee's Malt Shovel Brewery in Walsall. Lord's Short Acre site was closed down and brewing was concentrated at the larger Highgate site. Their activities attracted the attention of M&B who in turn took them over, along with their thirty-nine tied houses, in 1939.

By the 1950s there were three independent brewers in the Black Country, all of them Staffordshire based: M&B, W&DB and Butler's Springfield Brewery. In 1960 Butler's fell to M&B. Then there were two. M&B's chief rivalry was not with W&DB, but with Ansell's. There was talk of a merger, but it came to nothing. Both breweries now sought different roads. In 1961 Ansell's merged with Ind Coope and Tetley Walker to form Allied Lyons Plc, to become the largest brewing organisation in the world in terms of assets. That same year M&B merged with Bass Ratcliff & Gretton to become Bass, Mitchells & Butler Ltd. Six years later the company merged with Charrington United Breweries to become Bass-Charrington. The M&B name was lost, but the brewery was not. It was to become a flagship of the new company. They invested £11 million in expanding and upgrading Cape Hill. A further £10 million was spent on M&B tied houses. W&DB meanwhile continued to buy up houses and sell their less profitable houses to the smaller Black Country breweries. During the 1970s Ansell's was rocked by a number of strikes, culminating in the closure of the Aston Cross Brewery. M&B had seen off their chief West Midlands rival.

W&DB made an unsuccessful takeover bid for Birmingham's third brewery, Davenport's, in 1983, and again in 1986. This failure and the discovery of a listening device in the Davenport's boardroom soured any further relations. The rejection led W&DB to use the money in the acquisition of further new houses. By 1990 they owned 800 tied houses. During the 1990s Bank's mild was marketed simply as 'Bank's' allegedly – to fool the Home Counties' drinkers, who would not normally drink mild.

Bank's closed down Hanson's Brewery in Dudley in April 1992 and all production was moved to Wolverhampton. An opportunity for expansion presented itself in 1992 with the takeover of Cameron's of Hartlepool from Brent Walker, along with their fifty-one pubs. A later agreement with Marston's gave them the opportunity of selling their mild in Marston's houses in exchange for selling Pedigree in their own, thus, as they claimed, increasing choice for their customers. In 1999 W&DB bought out Marston's, increasing their tied houses to 1,763.

Amidst rumours that M&B might close down the Highgate Brewery, even though it was one of the most profitable in the Bass chain (it was producing over 75,000 gallons of mild a week), its management successfully negotiated a buy-out in 1995. As well as their Highgate Dark, which they renamed Highgate Dark Mild, the new company began brewing a new bitter called Saddlers (4% ABV). Dropping the word 'mild' had been a Bass marketing ploy to fool southerners, who saw mild as an old man's drink. Highgate pushed their mild to a more discerning drinker, looking for a change from the same old beers served up in the countless tied houses of the big chains. Highgate began expanding with the acquisition of four tied houses, and sought further outlets in the free-trade sector. Barely five years later, in the year 2000, Bass, renowned as brewers around the world, announced the sale of its beer production operations. Sale of their properties followed, including the Cape Hill Brewery, which was sold to Belgian brewing giants, Interbrew. What followed must have had the management at Highgate breathing a heavy sigh of relief that they had the foresight to buy up their brewery when they had. Meanwhile, following Government intervention over a seeming monopoly, Interbrew was forced to sell off part of their UK brewing concern, including Cape Hill, to the American Coors group. Cynically Coors closed down the Cape Hill Brewery at the end of 2002. All the pubs within the former Bass grouping, some 2,100 in number, were re-branded as Mitchell & Butler houses by their new management, Six Continents Retail. Without any sense of irony, the McAuliffe Group on behalf of Persimmon Homes, who had acquired the 56-acre site, decided to begin the demolition of the Cape Hill Brewery on the 127th anniversary of the laying of the first brick, 22 March 1878. Nature, however, decided differently. High winds postponed the start. All that remains of Cape Hill is the war memorial, the clock and the fire station. The site will eventually make way for 1,000 new homes. Highgate meanwhile was bought up by Aston Manor Brewery in 2000, by which time they had twelve tied houses and some 200 sales outlets. In 2002, Highgate announced its intention to brew Davenport's Bitter, following a gap of some ten years. How close it is to the original is debateable, even though the same recipe is being used.

Out of the blue, W&DB became the target of a hostile £453 million takeover bid from the Pubmaster Group. The bid was defeated by a very small majority, which set alarm bells ringing. W&DB was looking flabby. The company began to streamline itself. Cameron's was sold off to Castle Eden Brewery, and Marston's Mansfield Brewery was closed down, which pleased the Stock Market. Even so, at the end of 2002, Japanese investment bank Nomura, the UK's largest pub landlord, made enquiries as to a takeover. W&DB somehow managed to weather the storm, and consolidation became the new watchword. Two years later, in June 2004, the Black Country brewers bought Wizard Inns Ltd, a pub operator in southern England, for £8.9 million. At the beginning of December 2004, W&DB announced the takeover of Burtonwood Plc of Warrington, Cheshire, for £119 million. Burtonwood was a family run business that had merged with Thomas Hardy of Dorchester in 1998. They had 460 tied houses, mainly in the north-west, the Midlands and North Wales. The acquisition, it was claimed, would save W&DB £3 million a year in operating and distribution costs. At the moment W&DB are still major players.

FOUR

THE FUTURE IS SMALL

Well into the first decade of the twenty-first century we have seen the emergence of not just one, but four micro-breweries in the Black Country. In the autumn of 2004, four breweries emerged, indicating that brewing in the region is not dead, but vibrantly alive. Somehow because they are small they seem more in tune with the history of brewing in the Black Country. The rise of big breweries like Wolverhampton & Dudley Brewery and Mitchells & Butler's are an anomaly. Traditionally brewing occurred on a small scale – one pub, one brewery supplying it. At Netherton, Woodsetton and, of course, at Brierley Hill, there was some degree of expansion, with a brewery supplying a number of tied houses, but it was always on an intimate scale.

Of these new breweries, the Windsor Castle Brewery was established at Stourbridge Road, Lye and takes its name from the old Windsor Castle Inn in Oldbury. This was for many years run by the Sadler family, who are listed at Oldbury and Round Hill in the earlier list of local brewers. John Alexander, founder of the new brewery, learnt his trade from the last of the brewing Sadlers, John Caleb Nathaniel, and so in a way the tradition is kept up from earlier times. Alexander appointed his son, Christopher, as brewery manager. At present the brewery produces two beers, Sadler's 1900 Original Bitter, a malty bitter with a light hoppy aroma at an ABV of 4.5%, and Sadler's Jack's Ale, also a light hoppy beer with a suggestion of citrus. It has an ABV of 3.8%. Both beers were launched to the wider world at the Dudley Winter Ales Fayre on 25 November 2004.

The second of the new breweries is based at the Old Bull's Head, Gornall, a former home-brew pub in earlier days. The brewer is Guy Perry, formerly of the Sarah Hughes Brewery at Bilston Road, Sedgley. They have produced three beers: Pig on the Wall Mild at an ABV of 4.3%, Bradley's Finest Golden Bitter at 4.3%, and Fireside Bitter at 5%. They were also launched at the Dudley festival and are available over the border in Birmingham at the recently re-opened Wellington, in Bennetts Hill.

The Kinver Brewery, the third launched that autumn, is based at the rear of the Plough & Harrow in Kinver. It was established by Dave Kelly and Ian Davis. There are two beers to date, Caveman at 5.2%, and Over the Edge, at an interesting 7.6%. Finally there is the Toll End Brewery at the Waggon & Horses, Tipton. Their Cascade, at 4.2%, was also launched at the Dudley Winter Ales Fayre.

Windsor Castle Brewery beer labels.

There was one sad loss earlier that year – the Goldthorn Brewery Co. of Sunbeam Street, Wolverhampton, closed down. This five-barrel micro-brewery was established by brewer Paul Bradburn in April 2001, though test beers had been available for a short time before. In its three-year life the company brewed some thirty-two different beers and was a regular at most of the CAMRA beer festivals in the West Midlands. To compensate for this loss, the Beowulf Brewing Co. formerly of Waterloo Road, Yardley, in Birmingham, relocated to Chasewater Country Park, Brownhills, Walsall, in the autumn of 2003. Established by Phil and Claire Bennett in 1996, it was Birmingham's 'only independent brewery' as it styled itself. This micro-brewery, while it had no tied houses of its own, was supplying to over 300 free houses. In early 2003 it found itself in dispute with Birmingham City planners. A site in Sutton Coldfield was suggested, but this came to nothing. So it was that in the autumn of 2003 Beowulf relocated to Walsall, to Birmingham's loss and the Black Country's gain.

Will they all survive? Probably not, as Goldthorn has shown, but, given the history of brewing in the Black Country, this part of England does favour the small. It will not be easy. Brewing good, wholesome beer is only half of the story. Selling it in a competitive market is the real criteria. It will be hard work, and at times soul destroying. So at the end of the day it is up to the likes of you and me to drink as much of these new brews as we can, in order to make sure that they do not disappear. After all, if the breweries close down, that is the end of their beers. The big brewers in the past claimed, as with Allied Breweries when they closed down Ansell's at Aston in Birmingham, that they could replicate it at Burton. And could they? No they could not. It is up to us. Drink more Black Country beer, and drink it often!

A LIST OF
BLACK COUNTRY
BREWERIES

GEORGE ARNOLD & CO. LTD, Dartmouth Park Brewery, Dartmouth Street, West Bromwich.

Established as George Arnold & Co., New Street, West Bromwich in 1915. It registered as a company in 1919 before merging in 1924 with Henry Bates to form Arnold & Bates Ltd. The firm's address was given as 121 Spon Lane in 1928. The company was taken over by Darby's Brewery Ltd later that year.

BANKS & CO., Park Brewery, Wolverhampton.

A firm of maltsters established in 1840, they began brewing at Newbridge in 1874. In 1875 the company moved to the Park Brewery at Chapel Ash. See: Wolverhampton & Dudley Breweries.

HENRY BATES, Sponwell Brewery, Spon Lane, West Bromwich.

Formed in 1915, they merged with George Arnold & Co. in 1923 to become Arnold & Bates Ltd.

DANIEL BATHAM & SON LTD, Delph Brewery, Delph Road, Brierley Hill.

Situated behind The Vine, popularly known as the Bull & Bladder. Charlotte Batham, wife of the brewery founder, Daniel, first began brewing at their Cradley Heath home in about 1867. Daniel at this time was a colliery worker. In 1877 he gave up the mines and acquired a beerhouse in Netherton. He bought a second house, The White Horse, at 41 High Street, Cradley Heath, in 1882. Both home-brew houses remained in production until Batham moved into the Vine at Brierley Hill in 1905. This brewery should not be confused with the earlier Delph Brewery, which was situated a little way down the hill behind the Duke William Inn, now the Dock & Iron. By 1905, due to subsidence, most of the old brewery had fallen down. This had also effected The Vine, which was rebuilt in 1911-12 and renamed The Vine Hotel. Batham's prospered, and a number of tied houses were bought up. Daniel, who died in 1922, was succeeded by his son, Daniel Batham jr. Much of the day-to-day running of the business was left to his son, Arthur Joseph, who in 1941 registered the company as Batham & Son. After some consolidation, in 1923 the company began buying up pubs once more, and over the years accumulated nineteen tied houses. Daniel Batham jr died in 1939 and the company was obliged to sell off some of the houses to pay death duties. In 1951 Batham's

Right: *Banks' beer mat.*

Far right: *A Banks' advertisement.*

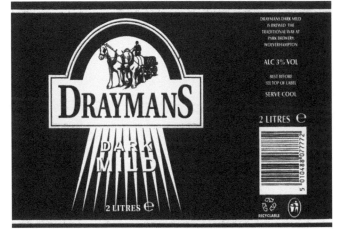

Above: *Banks' Coronation Old Ale, 1953.*

Right: *Drayman's mild, brewed for supermarkets.*

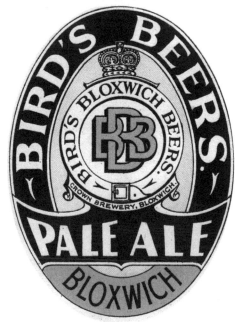

Above left: *Batham's beer mats.*

Above right: *Batham's Strong Ale.*

Right: *Bird's of Bloxwich.*

leased the Swan Inn in Chaddesley Corbett, and in so doing were obliged to radically change their beer production. Formerly, as in the rest of the Black Country, they had brewed a dark sweet mild. Such a beer proved unpopular in Chaddesley, so Batham's started to brew a pale ale, which was soon introduced into all their tied houses as Batham's Bitter. Talks were held with Holden's in 1953 with the intention of amalgamating the two companies, but this meant that all production would be moved to the larger site at Woodsetton, and the closure of the Delph Brewery. Nothing came of it in the end. However the family retained their links with the Holden's: one member is currently a director of the firm. Arthur Joseph Batham died in January 1974 and Daniel Bertram, better known as Arthur Batham, took over the reins. In the 1970s his son Tim trained as a brewer under the instruction of Philip Brown at W&DB. By 1990 Batham's had eight tied houses and had also moved into the free trade. In 1991 Batham's Best Bitter was voted winner in the best bitter category of the Great British Beer Festival in London. Batham's latest acquisition is the Y Giler Arms near Betws-y-Coed, Wales, which opened in March 2005. Batham's have gone international! The company produce a mild (3.5%), a bitter (4.5%) and a seasonal Christmas ale called XXX (6.5%).

JOHN BECKETT, Dudley Brewery, Ablewell House, Walsall.
Listed as a common brewer in the trade directory of 1839.

ARTHUR BEEBEE LTD, Malt Shovel Brewery, 130 Sandwell Street, Walsall.
Established by 1904 and registered as a company in January 1910. The address is alternatively given as 'Little London', Walsall. The company was taken over by Walsall Breweries Proprietary Ltd in 1924.

BENTS BREWERY CO. LTD, Cheapside, Wolverhampton and at New Brewery, Mount Road, Stone, Staffordshire.
In operation from 1915-24. In addition they also had premises at Hall Street, Bilston, in 1923. The company owned the Talbot Hotel in Digbeth, Walsall.

N.F. BIRD, Crown Brewery Ltd, 6 Leamore Lane, Bloxwich.
The Bird company was founded in 1864 and the Crown came down to the Birds by marriage through the maternal line. The pub and brewery were originally run by Samuel Birch, whose daughter married Josiah Brookes. Their daughter married James Bird, who developed the brewery. Bird was from a brewing family himself. His parents ran the Butts Inn and brewery in Walsall. Birds registered as a company in 1920 to acquire the brewery and its three tied houses. Some nine years later the company went into voluntary liquidation and James Bird invested the money in the Crown Brewery. The new company later acquired four tied houses and an off-licence. James' son, Norman, took over the Crown Brewery in 1939, as is shown in the trade directory of the day. The company acquired a further two public houses and a second off-licence. They also acted as bottlers for smaller breweries, including Booth's of Gornal Wood. N.F. Bird had taken over the business by 1960, but brewing ceased at the Crown Brewery in 1965. The firm continued as bottlers, but two years later the Crown Brewery was acquired by Ansell's, then part of Allied Breweries.

BLACK HORSE BREWERY See: Jack Downing.

BLOXWICH BREWERY CO. LTD, Elmore Green Road, Bloxwich.
Registered in August 1898, it built up a stable of forty-two tied houses. In 1924 the company entered into a partnership with the Town Brewery, Walsall, Highgate Brewery, and Pritchard's Brewery of Darlaston. A short-lived arrangement, Bloxwich Brewery was taken over by Butler's.

BOOTH, BRUFORD & CO., Steam Brewery, Market Place, Wolverhampton.
Founded in 1864, in 1872 the business became William Bruford & Co. No tied houses are known, so it would appear that the company was supplying the free trade. Bruford's closed in 1888.

THOMAS BOOTH, Red Lion Brewery, Market Place, Gornall Wood.
Former collier and Netherton publican Tommy Booth bought the pub and brewery in August 1935 from William Elwell for £3,650. Booth later rebuilt and extended the brewery to supply his seven local houses and the free trade. His beer was bottled by Bird's Crown Brewery of Bloxwich and Holden's of Woodsetton. The Red Lion and its brewery were taken over by Julia Hanson in November 1942. Three years previously, in 1939, Booth opened the Corbyns Hall Brewery in Tiled House Lane, Bromley, near Kingswinford. This was sold off following his death in 1952 and has subsequently been demolished.

WILLIAM BOWEN LTD, Cross Inn Brewery, 2 Oldbury Road, Greets Green, West Bromwich.
This was a private company registered in March 1927. No tied houses are recorded. It was taken over by Butler & Co. Ltd in 1945.

HENRY BRENNAND, Malthouse Brewery, Bratt Street, West Bromwich.
The brewery was designed by the London architectural firm of Inskipp & Mackenzie and opened in March 1903. Water for brewing was drawn from a bore-hole sunk to a depth of 350 feet. The brewery included a bottle-washing facility, van shed and stabling. It closed down in 1930.

BRITISH OAK BREWERY, Salop Street, Eve Hill, Dudley.
Originally a home-brew house established by 1855. It ceased brewing in 1898 and the pub was temporarily closed in the mid-1980s. Home-brewing was resumed there in May 1988 under Ian Skitt. Expanding, the company acquired a second tied house and fifteen free-trade outlets. Brewing ceased in 1997. The brewery formerly produced a mild (OG 1083), Eve'll Bitter (OG 1042) and Colonel Pickering Porter (O.G. 1046).

WILLIAM BUTLER & CO. LTD, Springfield Brewery, Grimstone Street, Wolverhampton.
The company was founded in 1840 as a retail brewers at a shop in John Street, Priestfield, also known as New Village, a suburb of Wolverhampton. Butler's first brewer was William Trueman, whom he employed in 1856. The brewery appears as Ettingshall Brewery in the directory of 1864. In 1871 Butler took on a partner, a Mr T. Russell, formerly of the Great Western Brewery. In 1873 the partners purchased a 7-acre site at Springfield, and the following year brewing was begun at the new Springfield Brewery. The brewery was designed to produce 400 barrels a week. Additional premises designed by London brewers' architects R.C. Sinclair were built in 1881-83. Sinclair designed a sixty-quarter brew-house tower, which has justifiably been Grade II listed. The new premises, connected to the old brewery, contained a sixty-quarter brewing plant and a 120-quarter malting. Production increased to 1,500 barrels a week. Two years later further improvements were undertaken, including the addition of a 100-quarter malting, stables and loose boxes for twenty-four horses. Butler's son, William Bailey Butler, had joined the company in 1877, and he was followed by Butler's other son, Edwin. Butler's became a registered company

BUTLER'S
WOLVERHAMPTON ALES
ON DRAUGHT,
Supplied at the Refreshment Bars,
ALSO
FINE BITTER ALE
(IN BOTTLE).

BUTLER'S
WOLVERHAMPTON
ALES
FOR FIFTY YEARS THE
STANDARD OF
STERLING QUALITY AND PURITY

Whenever you see the name BUTLER'S
on a Cask or Bottle of Beer you know
that it is the Best.

In the interest of your health it is
essential that you should drink beer
which is pure and good.

Butler's Ales are brewed from the finest
Malt and Hops that money can buy.

No Chemicals or Preservatives are used
in Butler's Beer: they are not required
when scrupulous cleanliness in the
Brewery is insisted upon.

Butler's Brewery at Springfields,
Wolverhampton, is celebrated for the
perfect conditions under which the beers
are brewed.

The reason why you should be careful
to ask for BUTLER'S is because it is a
PURE WHOLESOME ENGLISH BEER

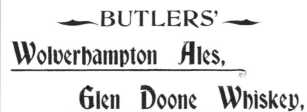

ＢＵＴＬＥＲＳ'
Wolverhampton Ales,
Glen Doone Whiskey,
AT ALL THEIR HOUSES.

Birmingham Offices :--Old Cross Street, off Coleshill St.

Top: *An early Butler's advertisement.*

Above and above right: *Butler's advertisements.*

Right: *Tommy Booth's Brewery.*

in April 1891, with capital of nearly £1 million. Butler sr died in 1893, and control of the brewery passed to his oldest son. In 1900 the company took over the brewery and licensed houses of J. Downing of Dudley, and those of the Bloxwich Brewery Co. Ltd, the Cannock Brewery and houses owned by William Blencowe & Co. Ltd, were acquired in 1925, and Eley's Stafford Brewery Ltd in 1928. In the post-war era, Butler's bought up the licensed premises of William Bowen Ltd of West Bromwich and the businesses of Thomas Oliver Ltd of West Bromwich, J.A. & A. Thompson Ltd of Oldbury and A.H. Clarke of Wellington, Shropshire. In 1946 James Pritchard & Son of Darlaston was taken over, and in 1947 Radcliff & Co. of Kidderminster. The company in its turn was taken over by M&B in 1960 and closed down some thirty-one years later on 2 August 1991. The brewery still operates as a sales and distribution centre.

BUTTS BREWERY CO. LTD, 46 Butts Street, Walsall.
Registered as a company in February 1920 to acquire the Butts Inn, with brewery attached, and its two tied houses. The company went into voluntary liquidation in October 1929.

JAMES CAHILL, Swan Brewery, St Matthew's Street, Heath Town, Wolverhampton.
Brewers and bottlers, established in 1907. Taken over by W. Butler & Co. Ltd in 1919.

CARR & CO. (HALESOWEN) LTD, Church Street, Halesowen.
Established in 1895 as Carr & Co. The brewery went into receivership in 1906, but continued trading as Carr & Co. (Halesowen) Brewery Ltd. The company is listed in Kelly's Directory of the Wine & Spirit Trades, Brewers and Maltsters for 1914 but was put up for sale in July 1915 following a court order.

CHAINMAKER BREWERY, Stourbridge. See: Premier Midland Ales.

JOHN CHAPMAN, West Bromwich Brewery, Churchfield, West Bromwich.
Common brewers, 1862-64.

C.H. CHATER & CO., Wolverhampton Brewery, Market Street, Wolverhampton.
Its tower brewery was designed by London architect Arthur Kinder in 1891. For its early history see: Old Wolverhampton Brewery Co.

CHESHIRE'S BREWERY LTD, Windmill Brewery, Windmill Lane, Smethwick.
The brewery took its name from an old windmill built around 1803 by William Croxhall and integrated into the brewery in 1886. Cheshire's registered as a company in October 1896 following the merger of Edward Cheshire and Benjamin Shakespeare. The company took over the six tied houses of Threlfall's Brewery Co. in 1898. Cheshire's were themselves taken over by M&B in 1913 and the brewery was closed down in the following year. Of the old windmill itself, it was demolished in 1949 and its working parts were moved to the Science Museum in London, where they were put on display.

JOSEPH COOPER & CO., The Brewery, New Road, Halesowen.
Established in 1895 as Joseph Cooper. The brewery registered as a limited company on 27 February 1900 to become J. Cooper & Co. Ltd. From 1914 the firm traded as Joseph Cooper & Son. It closed in early 1939.

HENRY COX & CO., Dudley New Brewery, Kate's Hill, Dudley.
Established in 1820 as Dudley New Brewery. It had changed its name by 1835 to the name of its proprietors to avoid confusion with the Dudley Brewery at Burnt Tree. The brewery was rebuilt as a ten-quarter in 1830-31. In September 1840 the partnership was dissolved however, and three months later the brewery and its contents were put up for auction. The auction particulars record the sale of 'Upwards of 250 barrels of Old Ale and Porter in barrels of 18 gallons each,' as well as '500 well seasoned Ale and Porter Barrels, namely 9, 18, 36 and 54 gallons'. The brewery itself was sold in June 1841. See also: Kate's Hill Brewery.

THOMAS CROSS, High Street, Bilston.
Maltster and common brewer, 1845.

Top left: *Cheshire's old brewery, c. 1960.*

Top right: *Office staff, Cheshire's Brewery.*

Above: *Arms Brewery.*

Left: *Cheshire's Windmill Ales.*

CRICKETERS' ARMS BREWERY, 10 King Street, Dudley.
Formerly known as the Horse & Jockey, it was renamed in the 1870s when Dudley Cricket Club made it their meeting place. The house was first recorded in 1822, when Samuel Holland was licensee. By 1913, when Henry Jones was landlord, the brewery had become a ten-quarter common brewery, brewing some 400 barrels a week. In October 1917 Jones established the Purity Bottling Stores in Claughton Road, Dudley, to bottle his widening range of beers. Later the company bottled beer for other smaller breweries. The Cricketers' Arms Brewery was taken over by the Diamond Brewery and closed down in 1922.

DARBY'S BREWERY LTD, Dunkirk Brewery, 6 Whitehall Road, Greets Green, West Bromwich.
Founded in 1894 by Charles Darby sr and registered as a company in 1923. In 1937 it bought out J.F.C. Jackson's Diamond Brewery Co. of Dudley. It was acquired along with its 100-plus tied houses by M&B in 1951. Brewing ceased shortly after.

THOMAS DARBY & SONS LTD, High Street, Old Hill, Cradley Heath.
Registered as a company in 1916, it fell victim to a hostile takeover bid by M&B.

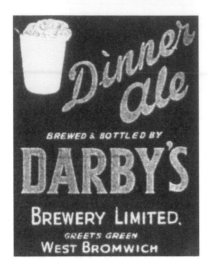

Above: *The Delph Brewery, Brierley Hill.*

Left: *Darby's Brewery, West Bromwich.*

JOSEPH HENRY DAVIES, Wheelwright's Arms, Castle Street, Netherton.
Sold to M&B along with its eleven pubs on 1 June 1942.

THOMAS DAWES, Church Street, Wolverhampton.
Listed as a common brewer in the trade directory of 1839. Gone by 1845.

DELPH BREWERY, Brierley Hill.
The brewery was established in 1876 by Benjamin Elwell on a 6-acre site. It was a twenty-five-quarter brewery, capable of producing 400 barrels a week. In addition to the brewery, a large wine and spirit store was established with extensive cellarage accommodation, together with a cooperage and stabling. By 1891 the company was employing a workforce of sixty. George Elwell had succeeded his father by then, and the company had over forty tied houses. The brewery amalgamated with Bucknall's Brewery of Kidderminster (founded in 1807), to form the Worcestershire Brewing & Malting Co., in 1896. The name was later changed to the Kidderminster Brewery Co. in 1905. The old brewery, surplus to requirements, was closed down. The Kidderminster Brewery Co. was taken over by W&DB in 1913. The Delph Brewery suffered serious subsidence due to mine workings below and was eventually demolished and the land cleared.

DIAMOND BREWERY, 19 Cromwell Street, Kate's Hill, Dudley.
Registered as the Diamond Brewery Co. Ltd by Joseph Plant in 1899. Two years later, in 1901, the company, having over-reached itself, was forced into bankruptcy. The brewery and its three pubs, its brewery tap, the Black Horse and the Loving Lamb, were put up for sale by auction. The brewery in its sales particulars was described as a five-quarter brewery producing 240 barrels a week. The brewery was bought up by the partnership of Hutchings & Jackson. See: J.F.C. Jackson.

J.A. DOWNES LTD, 16 New Street, Oldbury.
Registered as a company on 1 November 1894. Common brewers. No further details.

JACK DOWNING, Black Horse Brewery, Greystone Street, Dudley.

Dudley Brewery (above) and Enville Brewery (right).

Situated behind the Black Horse Inn, at 147 Upper High Street, the pub had been in existence since 1822, when Humphrey Hartle was landlord. The premises were bought by Jack Downing, formerly of the Leopard home-brew house, from the Diamond Brewery in June 1901. By 1912 he was brewing108 barrels a week. His most popular brew was 'Golden Sunshine'. Over the years he built up his tied houses to over twenty, before selling to W. Butler & Co. Ltd, in January 1923. The brewery closed down soon after, and the Black Horse was closed in 1958.

DUDLEY & DISTRICT BREWERIES LTD, St John Street, Netherton.
Registered as a company in October 1896 to acquire Rudge's High Street Brewery, Dudley; Salt's Brewery, Kate's Hill, Dudley; the Bull's Head Brewery, Netherton and the Talbot Brewery, Smethwick. The merger brought together 181 tied houses to the new company. Unfortunately the amalgamation collapsed after nine months amidst ill-feeling and rancour. The company was wound up on 10 June 1897. The Bull's Head and its outbuildings were bought up by the Netherton Bottling Co. The public house itself is now owned by W&DB.

THE DUDLEY BREWERY, Hall Street, Dudley.
Established by maltster George England in 1823. The trade directory of 1835 refers to the then company as George England & Son. By March 1845, son George Joseph was a full partner. By 1861 George sr had retired and the younger man had taken charge. Over-reaching himself in his expansion of the company, he was forced into bankruptcy in 1867 and was obliged to sell the company. It was bought by William Smith of the Netherton Brewery, Cinder Bank, Netherton. The next year though, in partnership with Rous John Cooper, England bought back the brewery. In 1871, Cooper bought out George England, and William Williams was installed as the company brewer. The company teetered on the point of bankruptcy for a number of years, until in 1881 it was taken over by George Thompson & Son. The Thompsons also owned the Victoria Brewery, which they had bought from John Dawes in 1880. The new company was renamed the Dudley & Victoria Breweries. In 1890 the company merged with Banks & Co. of Park Brewery, Wolverhampton, and C.C. Smith's Fox Brewery, also of Wolverhampton, to become the Wolverhampton & Dudley Breweries Ltd.

DUDLEY NEW BREWERY. See: Henry Cox & Co.

DUDLEY PORTER & ALE BREWERY CO., Burnt Tree, Dudley.
The company, comprising of local businessmen James Bourne, solicitor Joseph Royle, maltster Thomas Wainwright, a surgeon, and glass manufacturer Thomas Hawkes, was established in 1805

and the brewery was built in the following year. It was a ten-quarter brewery, producing between 250 and 300 barrels a week. It is first recorded in the trade directory of 1809. In 1828, following a series of buy-outs, the company changed its name to James Bourne, Cannon & Co., by which time it was known as the Old Dudley Brewery to distinguish it from Henry Cox's New Dudley Brewery at Kate's Hill. The company collapsed in 1835 and the brewery was taken over by Joshua Scholefield. In 1842, Scholefield took on two partners and the company was renamed Scholefield, Young & Stephen, with Thomas Dawes as manager. By 1846 the brewery had been renamed the Royal Brewery. In 1851 Samuel Allsopp & Sons of Burton-on-Trent bought up the brewery. This company later amalgamated with Tetley Walker and Ansell's to become Allied Breweries. Under Allsopp's, the Dudley Brewery was downgraded to a stores.

ENVILLE BREWERY, Cox Green, Enville, Staffordshire.

A micro-brewery established in April 1993 by Will Constantine-Cort in partnership with Mark Hill. Its head brewer was Richard Wintle, and production began at five barrels a week. The brewery is situated within a converted Victorian farm building, using the same water supply as the original village brewery, which closed in 1919. The brewery uses honey from its 250 beehives as the basis in the production of a number of its beers, based on a 120-year-old recipe. Its brews include Enville Bitter (3.8%), Low Gravity Mild (3.8%), Enville White (4.2%), Enville Ale (4.5%), Enville Gothic (5.2%), and Enville Porter (4.4%). Enville also serve Simpkiss Bitter (3.9%) from the original Brierley Hill recipe. The Cat at Enville serves as the brewery tap. In addition the company has some twenty or so trade outlets, mainly in the Black Country, but the Hog's Head, Newhall Street, Birmingham, regularly stocks Enville Ales.

ABRAHAM FISHER, Greet's Green, West Bromwich.

Maltster and common brewer, listed in the Staffordshire Directory of 1828. Sarah Fisher, Greet's Green Brewery, is listed for 1842, and Sarah and Louisa Maria Fisher appear in the directory at Greet's Green in 1845.

JESSE FISHER, Stoney Lane Brewery, Sandwell Road, West Bromwich.

The brewery was in existence from 1839 to 1842. It was later taken over by Heelas & Co.

FIVE WAYS BREWERY, Five Ways, Netherton.

The brewery evolved out of Thomas Penbury's home-brew house, the Five Ways Inn, which was in existence by 1832. It was established by John Rolinson, who was landlord in 1877. The brewery was built in the outbuildings behind the pub sometime after 1881. Extended, it became a fifteen-quarter brewery. In 1885 the firm became Robinson & Son when son Daniel joined the firm. The company began buying up public houses when they came up for sale. John Rolinson died at the age of seventy-four in January 1896. In July of that year the firm registered as a company. By then they were producing 850 barrels a week, and had six tied houses. In order to raise money for further expansion, Daniel Rolinson sold off three of the premises. In March 1899 the company was registered as John Robinson & Son Ltd, with a nominal capital of £100,000. At the time the brewery had built up its stable of tied houses, freehold public houses, beer-houses and off-licences to forty-seven. Edwin John Thompson of Wolverhampton & Dudley Breweries was invited to join the board. Daniel's extravagances led to financial crisis, and upon advice (or pressure) from the other directors, he severed his connections with the brewery in 1908. In August 1912 W&DB stepped in to buy the ailing brewery. The company continued to trade as John Robinson & Son Ltd up to 1925 when the company, with fifty-nine tied houses, went into voluntary liquidation.

JOHN FOLEY, Kate's Hill Brewery, Kate's Hill, Dudley.
Purchased by Foley in 1902, following the failure of Dudley District Breweries. Foley sold the premises to Thomas Plant in 1910, whereafter brewing ceased.

FOX'S BREWERY, Wolverhampton. See: Wolverhampton & Dudley Breweries.

GARRARD BROTHERS, Rowley Brewery, Long Lane, Blackheath.
The company formed part of North Worcestershire Breweries Ltd in 1896.

GOLDTHORN BREWERY, Imex Unit 60, Sunbeam Street, Wolverhampton.
A five-barrel micro-brewery set in the former Sunbeam car and motorcycle factory. Brewing started in January 2001 under brewer Paul Bradburn. Though it had no pubs of its own, it established some thirty outlets for the sale of its beer. The company's brews included Ge it Sum Ommer (ABV 3.8%) and Wulfrun Gold (4.3%). The brewery also produced seasonal beers; a winter ale called Deadly Nightshade (ABV 6%) and a spring-summer beer, Out of Darkness (4.3%). The Goldthorn Brewery & Co. ceased brewing in April 2004.

JULIA HANSON & SONS LTD, 89 High Street and Greystone Street, Dudley.
The company, established in 1847, was founded as a wine and spirit business in Priory Street, Dudley. In 1850 Thomas Hanson went into partnership with William Hughes, a maltster and publican, of Tower Street. The partnership was dissolved in 1864. Hanson died in 1870. His widow, Julia, took over the running of the business, which was relocated to Upper Tower Street. The company office was moved to 227 Market Place in 1882 to allow for the expansion of the business. Julia Hanson died in June 1894. Her sons, Thomas and William, began buying up pubs when they became available. They bought up the old Peacock Hotel and brewery in 1895. Hanson's registered as a company in October 1897 and their new brewery was built during that year. The brothers re-registered the company as Julia Hanson & Sons Ltd in 1902. By 1919 the company had over 100 tied houses and were also supplying the free trade. In May 1919 they took over Frederick Tandy's Brewery in Wood Street, Kidderminster, as well as its tied houses. In April 1934 they bought up Smith & William's Town Brewery in Round Oak, along with its fifty tied houses, including the Stew Pony & Foley Arms Hotel. W&DB had been buying shares in the company over a period of several years, and in 1943 acquired a controlling interest. The production of bitter (ABV 3.3%), a light but well-hopped beer, was moved to Wolverhampton. Hanson's were, however, permitted to brew their very distinctive medium dark and malty mild. The brewery continued in operation until 1991, when it was closed down.

HARMER & CO. (1912) LTD, Midland Brewery, Bilston Road, Wolverhampton.
Established in 1912. It was taken over by Ansell's post-First World War.

RICHARD ANDREW HARPER LTD, Hall Park Brewery, Wellington Road, Bilston.
The brewery was established in 1887. Extensions to the original small tower brewery were carried out in 1890, to the designs of Bristol and Worcestershire architects, Johnson, Charles & Son. The brewery was registered as a limited company in November 1916, following the death of Richard Harper. It was acquired by Ely's Stafford Brewery Ltd in 1924 and ceased brewing in 1928. The building was used as a shoe factory and is now a Hindu temple.

Harper's old Hall Park Brewery.

Davenport's is now brewed at Highgate.

HEELAS & CO., Burton Brewery, Sandwell Road, West Bromwich.
Founded in 1862, the company took over Jessie Fisher's Stoney Lane Brewery. In 1864 they are listed as 'East India pale ale, stout and porter brewers'. They apparently ceased trading the following year though.

HICKMAN & PULLEN LTD, High Bullen, Wednesbury.
Established in 1908 as a retail brewery, it had developed into a private company by 1915, and registered in July 1918 to acquire the business of Bernard Thomas Hickman and their five tied houses. The Receiver was called in, in January 1927, but the business continued trading until 1928, whereafter the company was wound up.

HIGHGATE (WALSALL) BREWERY CO. LTD, Sandymount Road, Walsall.
Registered as a company in August 1898, the brewery was built during the following year for J.A. Fletcher, son of a local wine and spirit merchant. It is now a DoE Grade II-listed building, being a very good example of an old tower brewery. Inside, some of the original equipment still survives. Fletcher's first brewer was Frederick Broadstock, who began brewing on 1 July 1899. During the First World War, Yardley & Ingrams' Brewery of Bloxwich was taken over, and production was transferred to Sandymount Road. In the early 1920s a bottling plant was established. As well as their own beers, Guinness was also bottled here for resale in the company's tied houses. Fletcher's also began blending wines and spirits. In 1924 J.A. Fletcher and John Lord of Short Acre, Walsall, combined together to form Walsall Breweries Proprietary Ltd, to take over the business of Arthur Beebee Ltd of the Malt Shovel Brewery in Walsall. All brewing was moved to Highgate.

In August 1939 the Highgate Brewery was taken over by M&B, along with the Walsall Breweries Proprietary Ltd and all the licensed premises of John Lord; some thirty tied houses. Under the subsequent Bass group, the Highgate (Walsall) Brewery was the smallest brewery in the company. There was always the threat hanging over it that it would eventually be closed down accordingly. Following a management buy-out in 1995 however, Highgate returned to the independent sector. Highgate Mild, sold to the south of England as Highgate Dark by Bass, was renamed Highgate Dark Mild. The new company also began brewing a bitter, which they called Saddlers, the first bitter brewed at Highgate in ninety-seven years. The company acquired a stable of four tied houses. In 2000, they were bought up by Aston Manor, of Aston, Birmingham. Highgate now has twelve tied houses, including the City Tavern, a restored Victorian public house off Broad Street in Birmingham, as well as some 200 outlets including Birmingham Airport. Highgate also supplies M&B houses with a variety of beers. Highgate brew Highgate Mild (ABV 3.2%), Highgate Old Ale (5.3%) and Saddlers Bitter (4.3%).

GEORGE HILL, 48 & 182 Park Street South, Blackenhall, Wolverhampton.
Established in 1915, most probably supplying the free trade. It was taken over by Atkinson's of Aston in Birmingham in 1919.

ROWLAND HILL, Angel Brewery, 37 Coventry Road, Stourbridge.
Founded by Hill, it had at least four tied houses. On 10 November 1911, then owned by H. & F. Kelly, the brewery and its tied houses were offered for sale by auction. The brewery eventually closed down in 1923.

THOMAS & STEPHEN HIPKINS, Old Swinford, Stourbridge.
Brewers and maltsters, 1845-54.

HIPKINS, MEEK & CO., Market Street, Wolverhampton.
Established in 1854, having evolved out of the above company. Short-lived, it had closed down by the following year.

HOLDEN'S BREWERY LTD, Hopden Brewery, George Street, Woodsetton.
The company was established by Edwin Alfred Holden at the Park Inn, Woodsetton, on 9 April 1920. Formerly Holden, along with his wife Blanche, daughter of Benjamin Round, had been licensees of a number of local pubs. In 1915 they purchased the freehold of the Park Inn

Above: *Holden's Brewery's tied houses.*

Right: *A pictorial advertisement for Holden's.*

at Woodsetton, a home-brew house. It was not until 1920 however that they moved into the Park from their previous pub, the Summer House. They succeeded in purchasing the maltings behind the pub, which were owned by Atkinson's of Aston. The building was converted into the Hopden Brewery. Holden's bought the Painters Arms in Coseley from Butler's Brewery, for their son, Teddy, a graduate in brewing from Birmingham University. He took over the business in 1938. Holden's registered as a company in 1964, with eleven tied houses. Edwin Holden, Teddy's son, joined the company in 1965 and ran it until his death in December 2002. Holden's have increased their tied houses to twenty-two, and also have a further ninety free-trade outlets. Holden's produce a Black Country Mild (3.7%), a bitter (3.9%), XB or LUCY B. (4.1%), Golden Glow (4.4%) and Special Bitter (5.1%). They also brew a stout (3.7%) as well as a seasonal old ale called XL (9%), with an OG of 1092. As well as draught beer, Holden's also produce a range of bottled beers, and bottled beers for other breweries.

HOLT, PLANT & DEAKIN, 480 Dudley Road, Wolverhampton.
A small brewery set up by Allied Breweries on 27 September 1984 at a cost of £100,000. It traded under the name Holt's and produced up to 5,000 gallons a week of an extra-strong ale called 'Entire'. Their mild and bitter were produced in Warrington 'to a special recipe created following extensive market research into the taste of Black Country drinkers', the advertising boys at Allied Breweries announced. The company initially supplied six tied houses, The Crosswells at Langley, The Fountain at Tipton, The Crown & Cushion at Ocker Hill, The Mount Pleasant at Sedgley, The Dudley Port at Dudley Port and The Posada at Wolverhampton. Eventually Holt's were provided by Allied, with a chain of forty-seven tied houses. The experiment over, Allied pulled the plug on Holt, Plant & Deakin and the brewery was closed down. It was purchased by Firkin, who reactivated the brewery to supply their local chain of tied houses. Holt, Plant & Deakin formerly brewed Entire (4.4%), Deakin's Downfall (5.9%) and Plant's Progress (5.9%).

HOME BREWERY (QUARRY BANK) LTD, Evers Street, Quarry Bank.
Founded by 1857, originally known as the Swan Brewery. It was bought out by Joseph Paskin Simpkiss in 1903 and registered as a company in December 1903 to acquire the business of J.P. Simpkiss. The brewery acquired twenty-three tied houses. Simpkiss lost control of the brewery after a lawsuit in 1916, and the brewery was taken over by former office manager, William Thomas Clewes. He did not possess the business acumen of Simpkiss, and the brewery went into decline. It closed in May 1921 and was demolished in February 1959. See: J.P. Simpkiss.

J. HORTON, Birmingham Street.
Ale and porter brewer, listed in the trade directory of 1839 as a common brewer.

SARAH HUGHES BREWERY, Beacon Hotel, 129, Bilston Road, Sedgley.
A former home-brew house, re-established in 1987, by John Hughes, Sarah's grandson, after lying idle for thirty years. It supplies the Beacon Hotel and a few other local outlets, as well as the free trade. The company also exports to the USA. The average brew can be up to nine barrels, with up to three brews a week. The Beacon was established by Abraham Carter in about 1865, and the business was maintained by his widow Nancy up to 1890, using outside brewers. The house had two more owners up to 1921, when it was bought by Sarah Hughes and her brewing partner, James Fellows. Sarah Hughes began brewing a strong-flavoured beer called Dark Ruby. Upon her death, her son Alfred Hughes took over. Brewing ceased in 1958. The present company have revived Dark Ruby Mild (ABV 6%) with an original gravity of 1058,

Right: *A Holt's beer mat.*

Below: *Holt, Plant & Deakin advertisement.*

WE'VE PUT THE H BACK IN THE LOCAL.

H stands for everything that made Black Country Pubs great. And H stands for Holts. Six traditional Black Country Pubs with all the values of the good old days.

'OSPITALITY.
When you walk into a Holts Pub, you're sure of a warm welcome and a smile.

And you'll have plenty of opportunities to show your skill at such timeless pub sports as darts and dominoes.

'OME COOKING.
The cooking in Holts Pubs makes a nice change. You'll love the good value traditional dishes and local specialities.

'OMELY ATMOSPHERE.
Holts Pubs are great places to have a drink with your mates. But whether you're with a crowd or not, you'll enjoy the friendly atmosphere.

Above all each Holts Pub has its own individual character.

Call into your local Holts Pub tonight.

It'll take you back a bit.

THE CROSSWELLS, LANGLEY
THE FOUNTAIN, TIPTON
THE CROWN & CUSHION, OCKER HILL
THE MOUNT PLEASANT, SEDGLEY
THE DUDLEY PORT, DUDLEY PORT
THE POSADA, WOLVERHAMPTON

'AND PULLED ALE.
Another welcome taste of the past is the full flavour of traditional hand pulled ale.

And you can expect every pint to be in the best possible condition because Holts landlords really care about their beer.

You'll also be able to enjoy a range of lagers and ciders, plus wines by the bottle or glass.

HOLTS

IT'S WHAT'S BEEN MISSING IN THE BLACK COUNTRY.

HOLT PLANT & DEAKIN · OLDBURY

Sarah Hughes' Brewery.

based on the 100-year-old recipe. They also brew a companion sweet ale called Sedgley Surprise (5%) and a sweet hoppy Pale Amber (4%).

J.F.C. JACKSON LTD, Diamond Brewery, 19 Cromwell Street, Kate's Hill, Dudley.
Originally registered as the Diamond Brewery Co. Ltd in 1899. It was taken over by the partnership of Hutchings & Jackson in 1901. John Jackson bought out his partner in 1912 and in 1916 the company re-registered as J.F.C. Jackson. The old brewery was retained. The Diamond Brewery and its fifteen tied houses were acquired by Darby's Brewery Ltd of West Bromwich in May 1937.

JOHNSON & PHIPPS LTD, 43 Lichfield Street, Wolverhampton.
Established in 1904 and registered as a company in 1912. It merged with J.P. Simpkiss & Son Ltd in 1956 to form J.P.S. Breweries Ltd. The brewery closed following the merger, and was later demolished when the area was redeveloped.

JONES & MATHER ALES LTD, North Street Trading Estate, Brierley Hill.
A short-lived brewery that started brewing in November 1996 and closed down at the end of 1997. The company produced seven beers; Dudley No.1 (ABV 3.8), Brewins Ale (4%), Brewins Bostin (4%), Telford's Tipple (4.4%), Crystal Cut (4.5%), Second Cut (4.8%) and Nobby's Goat (5%).

HARRY JONES, Cricketers Arms, 10 King Street, Dudley.
Originally a home-brew pub which expanded into a ten-quarter common brewery. The business was transferred to the Purity Bottling Stores, Claughton Road, in 1917. The brewery was then leased to the Diamond Brewery.

JONES & CO. BREWERY, 7 Snow Hill, Wolverhampton.
Established by the start of the **nineteenth** century, and perhaps the earliest common brewer listed for the Black Country, the company appears in the trade directory of 1802. William Jones, listed next door at No.9, probably the same man or a near relative, is also described as a brewer in the Wolverhampton Rate Book of that year.

JOHN JORDAN & CO., British Queen Breweries, Birmingham Road, Oldbury.
The business was founded in 1865 as a home-brew house. It expanded after the acquisition of a number of other public houses. Maltings were erected in Simpson Street, Oldbury, and in 1885 the company took over the old brewery in Cresswell Road, Langley, to expand production. The brewery and its tied houses were sold to M&B in 1920.

KATE'S HILL BREWERY, See: Samuel Salt.

CHARLES KING, Arden Grove Brewery, Langley.
A small business acquired by J. Nunneley & Co. Ltd of Burton-on-Trent in 1894.

J.W.J. KINGSTONE LTD, Summit Brewery, Great Arthur Street, Smethwick.
Founded in 1880 and registered as a company on 11 March 1924 to acquire the business of brewers carried on by J.W.J. Kingstone. Brewing ceased at the end of December 1927, and the brewery and its twenty tied houses were sold off at auction on 16 February 1928.

LAMSDALE & ECCLESTON LTD, Retreat Street, Wolverhampton.
A company that originated as a home-brew enterprise and expanded to become common brewers. They registered as a company in May 1905 to acquire the business of Lansdale & Eccleston, with their then eight tied houses. The company failed, and went into receivership on 23 October 1908. It was finally dissolved on 14 February 1917.

LASHFORD'S BREWERY, St Anne's Road, Willenhall.
Established by Jesse Lashford to supply the Springvale Tavern, on the corner of St Anne's Road and Spring Vale Street. The company expanded, buying up the New Inn at Darlaston Green. They acquired the Prince of Wales, King's Hill, Wednesbury for £880. The brewery also supplied the free trade in the area. Lashford's diversified into the leisure industry, buying up the old football ground at Willenhall and converting it into a greyhound track. Upon the death of her husband, Mrs Emily Lashford continued to run the business until 1930, when she sold the brewery, pubs and dog track to Truman, Hanbury & Buxton for £34,500.

JOHN LORD, Town Brewery, Short Acre Street, Walsall.
Situated next door to the Black Horse Inn, George Thomas brewed here up to 1895. In 1896 John Lord purchased the house and the land next to it. During the course of 1897-8, he built the four-storey Town Brewery. The firm originally traded as J. & H. Lord. An office was later established at 83 New Street, Walsall. In 1924 a consortium was established between them and the Highgate Brewery, Walsall, the Bloxwich Brewery Co., and Pritchard's Brewery of Darlaston in order to trade as a larger entity known as the Walsall Proprietary Ltd. In the process they took over the business of Arthur Beebee Ltd, the Malt Shovel Brewery in Sandwell Street, Walsall. Beer production was shifted to the Highgate Brewery. This growth, however, did not stop them from being taken over by M&B in August 1939.

Above: *The New Inn, a Lashford's house.*

Right: *Lord's Brewery, now an antiques centre.*

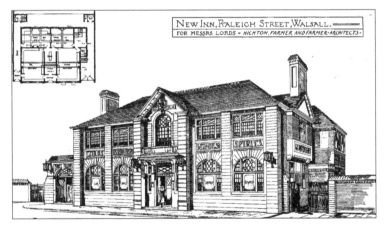

Left: *The New Inn, Walsall, a Lord's house.*

Below left: *A Millard's beer mat.*

Below: *The Gipsies' Tent, Millard's brewery tap.*

Lord's old brewery building is now used as an antiques centre. The Black Horse was unneccesarrily knocked down, and the site has not been redeveloped.

J. MADDOX, Cross Brewery, Oldbury Road, Tipton.
Listed in the Walsall Red Book of 1959. No other details available.

MALT HOUSE BREWERY. See: Henry Brennand.

MEAKIN & SONS, George Street, Walsall.
Common brewers, listed in the trade directory of 1839.

MIDLAND HOME BREWING CO. LTD, 15 Wisemore, Walsall.
The company was formed in February 1901 with a capital of £3,000 to acquire the business of J.F. Myatt. The new directors were A. Leary and J.F. Myatt.

BERT & DON MILLARD, Little Model Brewery, Stafford Street, Dudley.
The brewery tap was the Gypsies Tent in nearby Steppingstone Street. The pub was originally a home-brew house known as the Jolly Collier, and dated from 1804. It was acquired by George Thomas Millard in about 1870, and the name was changed to the Gypsies Tent. A brewery was built nearby for Millard, and opened in 1879. It was replaced by The Little Model Brewery in 1886. The new building was a three-quarter tower brewery of four storeys. In addition to a mild and a bitter, Millards also brewed an October Brew with a gravity of 1098. George Millard died in 1899. His widow, Harriett, took over control of the business. Their son Harry took over the business from his mother in 1914; Harriett died in 1921. Harry himself died at the age of seventy-seven in November 1957. The brewery and the Gypsies Tent passed to his two sons, Bert and Don Millard. Both men were bachelors. They were also tee-totallers. Brewing ceased in December 1961, owing to the Council's lack of assurances over the development of the area and increasing costs now faced by the brothers in keeping their pub up to date. The brothers bought in Bass for sale as well as cider, and the Gypsies Tent became known as a cider house. It eventually closed in 1980, and despite the redevelopment of the area, the pub remains looking as if come opening time it will again be dispensing drinks.

MITCHELL & BUTLER'S LTD, Cape Hill Brewery, Smethwick.
Henry Mitchell Sr established the business at the Crown Inn, Oldbury Road, Smethwick, in 1854. His son, Henry Mitchell jr, built the Crown Brewery next to the inn in 1866. He appears in the directory of 1872 as Henry Mitchell & Co. A new brewery, designed by the London-based architectural firm of Scamell & Colyer, was built at Cape Hill during 1878-79. By 1888 Cape Hill Brewery covered 14 acres and employed 271 people. The brewery had an annual output of 90,000 barrels. The company was incorporated in January 1888 to become Henry Mitchell & Co. (Ltd). The business continued to develop, and in 1898 an amalgamation took place between it and William Butler's Crown Brewery Ltd of Broad Street, Birmingham. The company became Mitchell & Butler's Ltd. All production was moved to the larger Smethwick site, which had the capacity for further expansion. In the period up to 1914, M&B swallowed up a number of smaller breweries, including Alfred Homer Ltd, the Vulcan Brewery, Aston, James Evans' Brewery in Perry Barr, and in 1913 Cheshire's Brewery, next door, in Smethwick, which was closed down in December 1914. Holder's Brewery Ltd and the Midland Brewery were acquired in 1919, and the Highgate (Walsall) Brewery Co. Ltd in 1939, along with its thirty-nine tied houses. In the

Post-war M&B advertisement.

Railway sidings at Cape Hill Brewery, 1929.

Above: *M&B All-Bright Ale. and M&B Sweet Stout.*

Right: *Brewing coppers at Cape Hill, 1929.*

immediate post-war era, as well as brewing mild and bitter, Head Brewer H.J. Cox also produced a number of bottled beers for M&B, including Cape Ale, Export Pale Ale, Family Ale and Sam Brown, a brown ale. The company was also looking to expand. Darby's Brewery Ltd, Whitehall Road, West Bromwich was taken over in 1951. Atkinson's Brewery of Queen's Road, Aston, was bought up in 1959, and W. Butler & Co.'s Springfield Brewery in 1960. In 1961 M&B merged with Bass, Ratcliff & Gretton Ltd of Burton-on-Trent to form Bass, Mitchell & Butler's Ltd. Six years later the new company further merged with Charrington United Breweries to become Bass-Charrington, with 11,500 tied public houses and hotels. An £11 million expansion programme was undertaken at Cape Hill, and a further £10 million was spent upgrading and refurbishing their pubs in the following twenty years. In the year 2000, following a shift in company policy, Bass announced the sale of its beer production operations. Such a sale included their breweries at Cape Hill and Burton-on-Trent. In June 2000 the brewery was taken over by Belgian brewers, Interbrew. Following Government intervention concerning monopolies, Interbrew were forced to sell off part of their British brewing concerns. These were bought up by American brewing giants, Coors. In 2002, Cape Hill Brewery was closed down. M&B pubs continue to operate under their old brewery name, through Bass subsidiary company Six Continents Retail. Before the Bass sell off, Cape Hill brewed a mild (ABV3.3%), a bitter, Brew XI (3.9%), and Charrington IPA (ABV 3.4%), not generally available in the Black Country.

Above: *Myatt's Old Time Strong Ale.*

Left: *The Royal Oak, a Myatt's house.*

FRANK MYATT LTD, West End Brewery, Raglan Street, Wolverhampton.
Founded in 1900 at the Cross Keys public house by John Francis and registered as the Midlands Home-Brewing Co. Ltd in December of that year. Due to increased productivity, the company moved to the Albany Brewery. The name of the company was changed to Frank Myatt & Co. Ltd in August 1902. The business was sold to Eley's (Stafford) Brewery Ltd in 1909. Myatt then went on to establish another home-brew house at the West End Inn. He was able to buy up the West Midlands' houses of the Manchester Brewery Co. Ltd when they became available. In June 1920 Frank Myatt Ltd was registered as a limited liability company to acquire the business of Frank Myatt & Co. Ltd, West End Brewery, Raglan Street, Wolverhampton, and the Old Wolverhampton Breweries Ltd, Market Street, Wolverhampton, along with its 124 tied houses. In February 1927, Wolverhampton Corporation acquired by compulsory purchase the brewery block in Market Street with 1,196 square yards of land attached, valued at £33,391. In addition, the Council also purchased the Red Cow and the Alhambra, two of Myatt's houses situated nearby, as part of an improvement scheme in the town centre, attached to the building of the Birmingham to Wolverhampton Road. Left without a brewery, Myatt's remaining houses were taken over by the Holt Brewery of Aston later that year.

J.F. MYATT, Albany Brewery, Wolverhampton.
See the above. The business was acquired by the Midland Home Brewing Co. Ltd.

NETHERTON BOTTLING CO. LTD, Simms Lane, Netherton.
The company was registered in April 1925 with a capital of £2,000 to acquire the business of brewers, beer bottlers and wine and spirit merchants carried on by F.H. Billington and B. Billington at the Queen's Head, Simms Lane, Netherton.

NETHERTON OLD BREWERY, 167 High Street, Netherton.
The brewery was developed around 1830 by former miner Thomas Hotchkiss, who also carried on the trade of maltster. Upon his death his sons, Thomas and William, carried on the business. The brewery is first recorded as 'The Old Brewery, Ale & Porter Brewers, Sweet Turf,' in the Worcestershire directory of 1860. While it does not appear that the brewery owned any tied houses, they did supply the free trade. The family sold the brewery and adjoining tap house,

the Castle Inn, to the North Worcestershire Breweries, who were themselves taken over by Wolverhampton & Dudley Breweries. In 1921 the Castle Hotel as it had become, then owned by Plants, was taken over by Ansell's. The site was cleared and a new pub, the Mash Tun, was built to replace it.

HERBERT NEWNAM & SONS LTD, 101 Pedmore Road, Lye, Stourbridge.
Situated next door to the Seven Stars public house. Established by 1895. During the 1930s the brewery produced a strong, very pale ale, at an OG of 1102. This beer was stronger even than Bass' 'No.1 Barley Wine' (OG 1101). This beer was sold under the name 'Premier Cuvee' with the advertising gimmick, 'Treat it with respect'. Three pints was enough to floor most men. Newman's was acquired by Wolverhampton & Dudley Breweries in 1960.

NORTH WORCESTERSHIRE BREWERIES LTD, Stourbridge Brewery, Duke Street, Stourbridge.
Registered as a company in May 1886 to acquire and amalgamate the Stourbridge Brewery, the Rowley Brewery, Black Heath, the White Swan Brewery, Oldbury and the Round Oak Brewery, Brierley Hill, with a combined total of 135 tied houses. Brewing thereafter was concentrated at the Stourbridge site. The Round Oak Brewery was sold to Elwell Williams in 1897 and renamed the Town Brewery. The brewery was reconstructed and enlarged after a fire, by architects Johnson Charles & Son in 1897. The firm was taken over by Wolverhampton & Dudley Breweries in 1910.

OLD LION BREWERY LTD, 140 Park Lane West, Tipton.
Registered as a company in April 1895. Taken over by Peter Walker (Warrington & Bolton) Ltd in March 1898.

OLD SWAN INN BREWERY, Halesowen Road, Netherton.
Known nationally and affectionately as Ma Pardoe's, after Doris Pardoe, longtime landlady of the Swan. The house has its origins in the early **nineteenth** century. Then a beer-house, its first known licensee was Thomas James in 1835. He was succeeded by Joseph Turner, who held the premises from 1841 to 1850, Thomas James, 1850-51, John Roper 1851-61, John Northall 1861-62, John

The Old Swan, Netherton.   *The Old Swan Brewery.*

Young 1862-71, under whom the terrace of houses that includes the Old White Swan was rebuilt. George Baker was landlord from 1871-84, Edward Evans 1884-92, John Andrew Harris 1892-98, Albert Harvey 1898-1901, William Chiltern 1901-12 (with Zachariah Marsh as manager from 1909-12), Albert Lyndon 1912-24, Clifford Harris Pearson 1924-27, Thomas Hartshorne jr 1827-28 and Harry Brown 1928-32.

In 1932 Frederick Pardoe, late of the British Oak, Sweet Turf, Netherton, took up the tenancy of the Old Swan. Ben Cole remained as brewer at the Old Swan. He was succeeded by Solomon Cooksey, and he in turn by his son George, who remained as brewer up to September 1988. Meanwhile licensee, Fred Pardoe, expanded the business by buying up properties. He bought the Gladstone Arms at Audnam, near Brierley Hill, the White Swan in Holland Street, Dudley, now rebuilt in 1970, and an off-licence in New Street, Dudley. Frederick Pardoe died in 1952 and his widow Doris took over the license. In 1964 she purchased the freehold of the property, thus ensuring its status as an independent brewery. It achieved national honour during the 1970s as being one of only four surviving home-brew houses in England. With the promotion of CAMRA, 'Ma Pardoe's' became a must for beer drinkers from all over the country. Wolverhampton & Dudley Breweries made tentative offers, but these were declined. Doris Pardoe died on 1 April 1984. Daughter Brenda and husband Sid Allport took over the license. Within a year though the Old Swan was put up for sale. Anxious that it should retain its independent status, CAMRA and Mercia Venture Capital Ltd formed Netherton Ales Plc. and purchased the pub and its brewery. To make it a more viable proposition, the new company extended the premises to the side and rear. The new company, comprised of enthusiastic amateurs, soon found itself in difficulty. Hoskins of Leicester, a small independent brewery, stepped in and bought up the Old Swan in 1987. Premier Midland Ales, a Stourbridge-based group, took over the Old Swan in 1990 in an amalgamation of a number of other independent Black Country and Coalbrookdale pubs. Following a merger with the Wiltshire Brewery of Tisbury, a new company, County Inns, was formed and brewing at the Old Swan Brewery was resumed in 1991 after a lapse of one year.

OLD WOLVERHAMPTON BREWERIES LTD, South Staffs. Brewery, Market Street, Princes End, Wolverhampton.
The company was founded in 1871 by Reuben Turner, though Turner had been brewing since 1864. The brewery was registered as a company in April 1910 to acquire J. & J. Yardley and I. Yardley & Sons of Bloxwich (then in liquidation), and the Darlaston and the South Staffordshire Brewery Co. Ltd of Wolverhampton, as well as 130 tied houses. The company was taken over by Frank Myatt Ltd in 1919. Listed in the directories as the Old Wolverhampton Breweries in 1924, it was later acquired by Ansell's.

OLD WOLVERHAMPTON BREWERY CO., Market Street, Wolverhampton.
Formed in January 1897 to acquire the properties of the Wolverhampton Brewery (late C.H. Chater & Co.) in Market Street, the Hall Park Brewery (late Harper's) at Bilston, along with 140 tied houses. See also: Frank Myatt Ltd.

THOMAS OLIVER LTD, Sandwell Brewery, Walsall Street, West Bromwich.
Established in 1904. Oliver registered his Throstle Ales trademark in 1912. The business was taken over by W. Butler & Co. Ltd in 1945.

W. OLIVER & SONS OF CRADLEY LTD, Talbot Brewery, Colley Gate, Cradley.

Registered as a company in December 1915. The company was taken over by Darby's Brewery Ltd of West Bromwich in 1937.

WILLIAM ONSLOW, Netherton New Brewery, Hill Street, Netherton.
Originally a home-brew house, the company was founded in August 1875. It later merged with John Robinson & Son Ltd. See: Five Ways Brewery.

PARDOE'S, Old Swan Brewery, 14 Halesowen Road, Netherton.
See: Old Swan Brewery.

PENN BREWERY CO. LTD, Penn Wood Common Brewery, Wolverhampton.
Run by Thomas Fox Allin from 1872. It registered as a company in October 1887 to acquire the business of Williams & Empson. It became Hall & Haden's Penn Brewery in 1888. V.E. Heathcote-Hacker was appointed manager of the new company. In 1897 Penn Brewery took over the William Harper Brewery of Dudley. They were themselves taken over later that year by the Burton Brewery Co. The brewery was later sold to the Wolverhampton District Brewery Ltd in August 1899. Later acquired by Ansell's.

THOMAS PLANT & CO. LTD, Steam Brewery, Netherton.
Founded by William Round in 1837, it was acquired by Thomas Plant in 1875. By 1881 he was producing a range of sixteen draught beers. Thomas Plant died in 1895 and, with no son to succeed, the business passed into the hands of his executors. Plant's was registered as a company in 1901 and John Shaw was appointed brewery manager. The firm was taken over by Hereford & Tredegar Brewery Ltd in June 1912 and closed in 1914. It was re-opened the following year and Shaw was re-appointed to run the brewery. Never completely financially viable, Plant & Co. was taken over by Ansell's Ltd in 1936, along with its sixty-three tied houses. The brewery was closed down again in 1947. During the 1950s the Steam Brewery and its malt houses and stores were demolished and houses were built on the site, just off St John's Street.

PREMIER MIDLAND ALES, Mill Race Lane, Stourbridge Industrial Estate.
Founded in 1988, eventually acquiring a stable of seven tied houses. It merged with Pitfield Brewery, London, to form Pitfield's Premier Brewing Co., concentrating brewing at Stourbridge. The company later went into voluntary liquidation and in 1991 the brewery and its houses were taken over by United Breweries of India. The operation crashed later that year and the brewery closed.

ARTHUR JAMES PRICE, Lewisham Brewery, 43-45 High Street, West Bromwich.
The company was founded in 1887. Following Price's death, his executors sold the brewery in May 1908, along with its forty tied houses, to Holder's Brewery of Birmingham.

JAMES PRITCHARD & SON, Darlaston Brewery, Church Street, Darlaston.
The company was founded in 1899, a year after Lord's Town Brewery in Walsall. In 1924 these two breweries, plus the Highgate Brewery, Walsall, and Bloxwich Brewery, formed a short-lived consortium. This broke up when the Town Brewery and Highgate were taken over by M&B in 1939 and Pritchard's was forced to close.

QUEEN'S CROSS BREWERY (also known as The Dudley Hop Ale Brewery), Queen's Cross, Dudley.

Above: The Queen's Brewery, Dudley.

Right: Matthew Smith's Queen's Brewery.

MATTHEW SMITH,

Queen's Cross Brewery,

DUDLEY.

NOTED

PALE & MILD ALES,

❖ **Stout and Porter.** ❖

THE TRADE AND FAMILIES SUPPLIED.

FINE FAMILY ALE,

AT

ONE SHILLING PER GALLON.

A ten-quarter three-storey brewery opened by Matthew Smith prior to 1878. The brewery was capable of producing up to 500 barrels a week. In addition to the obligatory mild, bitter, stout and porter, the firm also produced Dudley Hop-Ale, a non-alcoholic sweet bottled beer selling at 2d a half pint. The firm acquired seven tied houses, mainly in Dudley itself, with one at Kingswinford. Smith died in 1914, at the age of eighty-two, and the business was placed in the hands of his executors, who ran it until 1917. It was then bought by H. & B. Woodhouse, formerly of the Alma Brewery in Hall Street. They retained the premises until 1934 when they ceased brewing. The brewery was leased to Francis Billingham of Netherton. In the post-war era it was used as the Midland stores of Cardiff Brewers, William Hancock. In 1950 the Lamp and its brewery buildings were taken over by Daniel Batham. The pub is still a Batham's house, while the brewery has been renovated and converted to other purposes.

THOMAS ROBINSON, Smethwick.
Brewer and maltster of Smethwick, listed in the trade directory of 1839. No further directory listings after this date.

ROGERS & CALCUTT LTD, 70 Steelhouse Lane, Wolverhampton.
Registered as a company in 1920 to acquire the business of John Rogers. The company was taken over by Atkinson's Brewery of Aston, Birmingham.

JOHN ROLINSON & SON LTD. See: Five Ways Brewery.

THOMAS ROUND, Lion Brewery, Park Lane West, Tipton Green.
Brewers and maltsters from 1861 to 1878. The Old Lion Brewery was acquired by Ansell's.

WILLIAM ROUND, Steam Brewery, St John's Street, Netherton.
William Round first began brewing beer in 1837 from his Cottage Spring public house. Upon his death in 1854 his son Samuel took over the business. In November 1860 he bought additional land behind the pub and, following a further purchase in March 1861, built an eight-quarter

small tower brewery which included a four horse-power steam engine with outside boiler, thus giving the brewery its name. The brewery was capable of producing up to forty barrels a day. The company acquired a number of tied houses, and also sold to the free trade. Samuel died in December 1874, leaving the brewery and pubs to his son, Jabez Edward Round. At Christmas time the following year, Jabez Round sold off his inheritance to Thomas Plant of Brierley Hill.

RUSHALL BREWERY LTD, Rushall.

Registered as a company in March 1905 as the Old Rushall Brewery Ltd. The new company registered as the above in the following April. They ceased brewing in December 1915. The premises were later taken over by Walsall & District Clubs Brewery, formed in 1920.

THOMAS RUSSELL LTD, Great Western Brewery, Great Western Street, Wolverhampton.

Founded in 1877, the company registered in 1900 as a limited liability company. It was taken over by W. Butler & Co. Ltd in 1932, along with its fourteen tied houses. The brewery was then closed. Records relating to the company are available in the Wolverhampton & Local Studies Library.

SAMUEL SALT, Kate's Hill Brewery, 60 St John's Street, Dudley.

This appears to be the old Cox & Co.'s Dudley New Brewery. Salt was licensee of the Malt Shovel, which fronted the old brewery in 1861. During the 1870s Salt expanded into malting, and his brewing may originate from this time. He is listed in the directories from 1880 to 1888. Dead by 1891, the brewery was in the hands of his executors, with Arthur Edward Lloyd as manager. As well as supplying their own tied houses, the brewery also supplied the free trade. Sam Timmins of the Cape of Good Hope at Tipton, was one of his customers. Sam's son, Samuel jr, having taken over the business, himself died in 1895, aged thirty-one. The brewery was leased to John Foley in 1896. In 1902 it was put up for sale. The particulars refer to it as a ten-quarter, with two malt houses attached, of fifteen and ten quarters. Along with the brewery, its five tied houses were also offered for sale. Foley succeeded in buying the brewery, and he and his wife, Annie, traded from here until 1910. That year it was bought by Thomas Plant, whereafter brewing was discontinued. The property was taken over by Ansell's.

JOSHUA SCHOLEFIELD & CO., Dudley Port.

A common brewer, listed in the trade directory for 1839. See also: Dudley Old Brewery, which was run by his son, Joshua jr.

JOHN SEEDHOUSE & SONS LTD, Seven Stars Brewery, High Street, Prince's End, Tipton.

Originally a home-brew house based at the Seven Stars Hotel, the company was founded by John Seedhouse in 1858. An extensive brewery was later built as a chain of tied houses was acquired. The company later began to supply beer to many working men's clubs in the district. It became a private company, registering in 1912. The company was reconstituted in 1932. Seedhouse's ceased brewing in 1955 and their fifteen tied houses were sold to Wolverhampton & Dudley Breweries. The company continued to operate after this period as a wine and spirit merchants.

SHELTON BREWERY, Queen's Arms, Graisley Row, Wolverhampton.

A home-brew house that expanded its production to supply other outlets. The brewery was the town's last traditional home-brew house. Built prior to 1869, the brewery also supplied the free

An advertisement for John Seedhouse, 1950.

A Seedhouse beer label.

trade. The brewery also bottled beer for themselves and neighbouring home-brew houses. The establishment was run pre-war by Charles Hilton jr. Upon his death the brewery and pub were taken over by his brewer, a Mr Shelton. Brewing ceased in December 1959 and the house was bought up by Ansell's in 1964. It is now owned by Burtonwood.

SHOWELL'S BREWERY CO. LTD, Crosswells Brewery, Langley, Oldbury, with stores at Kingston Buildings, Crescent Wharf, and offices at Great Charles Street, Birmingham.

The company was founded by Walter Showell, a Birmingham man. A former chemist and later bakery and beerhouse proprietor, he opened a small brewery in Simpson Street, Oldbury. Proving a success, he secured a large piece of land, including the historical Crosswell Spring near the Great Western Railway at Langley. Here he built his Crosswell Brewery, which opened in 1870. A second ninety-quarter brewery was added, to a design by Davison, Inskipp & Mackenzie, in 1884. They designed further extensions in 1885. In 1884 the company became Walter Showell & Son. It was registered as a limited liability company in March 1887, when Walter Showell handed over control to his son Charles. By 1890 the company had forty tied houses in Birmingham and, with the acquisition of the Brewers' Investment Corporation Ltd in 1894, gained a further forty houses within the city. Showell's took over Taylor's Hockley Brewery, Birmingham, and Sarah Marsland's Brookfield Brewery of Stockport, in 1890. The company's name was changed to Showell's Brewery Co. Ltd in 1894, and their head office was moved to Great Charles Street, Birmingham, in the heart of the financial district. Extensive storage and bottling facilities were also acquired near the Crescent, off Broad Street, with a frontage onto the Birmingham & Wolverhampton Canal. Two years later Showell's sold off their established second brewery at Brookfield Brewery, Stockport, for £250,000, in what appears to have been a financial crisis. Their London tied houses were sold to Refell's Bexley Brewery in 1899. In 1914 Showell's was taken over by Samuel Allsopp Ltd of Burton-on-Trent, along with their 194 tied local houses and thirty off-licences. Brewing was discontinued at Langley. All company properties were leased to Ind Coope and Allsopp Ltd. The Crosswell

Trade Mark.	Name, Address, and Calling of Applicant.	Class of Goods.	Description of Goods.	Number given by Registrar.	Date of Application received.
	WALTER SHOWELL, Crosswells Brewery, Langley Green, near Oldbury, Worcestershire; Brewer.	43	Ales, Porter, Stout, Wines, and Spirits.	22,049	5th Mar. 1880

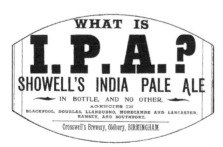

Top: *Showell's registered trademark.*

Centre: *A Showell's beer label.*

Above: *Twyford's 'What is IPA' campaign.*

Right: *Harry Twyford, a Showell's director.*

Above left: *Simpkiss' Nut Brown Ale.*

Above centre: *Simpkiss' Special Bitter.*

Above right: *Simpkiss' Old Ale.*

Left: *A Simpkiss' beer mat.*

site was eventually sold off in 1938. The company, then known as Ind Coope (South Midlands) Ltd, went into liquidation in 1966.

J.P. SIMPKISS & SON LTD, Dennis Brewery, Brettell Lane, Brierley Hill.
The business was founded as a home-brew house at the Potter Arms, at the Delph, by William Simpkiss in 1854. William's son, William Henry, bought the Royal Oak in Round Hill in 1869 and later built a brewery on the land behind. In 1896 the North Worcestershire Breweries bought him out for £20,000 and he retired. In 1903, financed by his father, William Henry, Joseph Paskin Simpkiss bought the Swan Brewery in Evers Street, Quarry Bank. He renamed it the Home Brewery. The company acquired twenty-three tied houses over the years. At the outbreak of the First World War they were producing over 300 barrels a week. Simpkiss lost the brewery in 1916 in a curious court case which revealed that, for some strange reason, he had signed over the business in favour of his brewery manager, who promptly ousted him. Simpkiss became a travelling representative until 1919, when he had raised enough money to buy the Foley Arms in Brettell Lane, Brierley Hill. Joseph Paskin Simpkiss established his new business on 26 August 1919, brewing from a room that was later to become the pub lounge. In 1934 a new brewery was built on a site just behind the Foley Arms. The company was registered in November 1938. Simpkiss merged with Johnson & Phipps Ltd of Wolverhampton in 1955 and the new firm became J.P.S. Breweries Ltd. In 1977 the company name reverted to J.P. Simpkiss & Son Ltd. Dennis Simpkiss died in 1981. A takeover bid from Greenhall-Whitley in 1984 was rejected, but in July of the following year the company agreed to the sale of the brewery and its fifteen tied houses. Greenall's, with unbelievable cynicism, closed down the brewery very soon after, with a loss of twenty jobs. Simpkiss had formerly brewed two regular beers: Simpkiss Bitter (OG 1037) and Simpkiss Old Ale (OG 1050).

ARTHUR JOEL SMITH, Lion Brewery, Park Lane West, Tipton.
'Celebrated strong and mild ales, porter & stout, maltster & hop dealer, wine & spirit merchant…
Established 1791.' This was the regular entry in Kelly's Directory of Staffordshire during the
1880s. The firm later became Smith & Barton.

MATTHEW SMITH, Queen's Cross Brewery, Queen's Cross, Dudley.
Built about 1873, it was sold to H.B. Woodhouse in 1917. Brewing ceased in 1934.

SMITH & WILLIAMS, See: Town Brewery, Round Oak.

SMITHFIELD BREWERY, Market Street, Wolverhampton.
Established by Reuben Turner, who ran it from 1861 to 1878. The firm became Turner &
Wilkins in 1883. The brewery was taken over by Ansell's.

SOUTH STAFFORDSHIRE BREWERY CO. LTD, Market Street, Wolverhampton.
Founded in 1887 with Robert Muras as Secretary. See: Old Wolverhampton Breweries.

STAFFORDSHIRE UNITED BREWERIES LTD, Market Street, Wolverhampton.
Registered as a company in October 1896. It was dissolved on 22 December 1903.

TOLL END CROWN BREWERY CO. LTD, 40 Toll End, Tipton.
Registered as a company on 22 November 1879 to carry on the business of brewers. The
company was wound up soon after, possibly before they had ever brewed any beer.

TOWN BREWERY, Round Oak.
Originally a home-brew house called the Royal Oak Inn, dating from pre-1828. Henry
Husselbee was licensee in 1834. The inn was bought by J.P. Simpkiss in 1869 and he built the
brewery behind. By 1890 the company had three ties houses and were brewing between 250
and 350 barrels a week. In 1896 the North Worcestershire Breweries bought the brewery and its
tied houses for £20,000. The brewery was sold the following year to E.H. Smith, Elwell Williams
and partners. The new company renamed the brewery the Town Brewery and rebuilt the stables
and bottle store, as well as making general improvements to the brewery itself. The company
evolved into Smith & Williams in 1916. Having lost control of the Home Brewery, J.P. Simpkiss
worked briefly for the firm as a commercial traveller. In 1834 Julia Hanson & Sons bought up
the brewery and its sixty tied houses, including the Stewpony and Foley Arms, for £120,000.
The brewery was closed down soon after.

TOWN BREWERY, Walsall, See: John Lord.

TWIST'S BREWERY, White Horse Brewery, 92 Wolverhampton Street, Walsall.
Established by George S. Twist, it is listed in the trade directories from 1928. It was registered
as a private company in April 1943. Twist's Brewery was taken over by Atkinson's of Aston,
Birmingham, in 1950, along with its twenty public houses.

VICTORIA BREWERY, 30 Hall Street, Dudley.
Established by John Dawes, licensee of the Lamp Tavern, in Hall Street. He leased the premises

Above left: *Walsall & District Clubs Old Ale.*

Above right: *Walsall & District Clubs bitter.*

behind his pub, formerly used by maltsters, William and Luke Jukes, and began brewing in 1873. Dawes further leased the Seven Stars in the High Street and bought two further beerhouses from which to sell his beer. It seems that he over-reached himself financially, because in 1880 he became bankrupt. The properties were bought by George Thompson & Son, prime movers in the future amalgamation that brought about Wolverhampton & Dudley Breweries.

JOHN WALL, Duke Street, Stourbridge.
Founded by John Wall in 1863, and expanding from a home-brew house into a common brewers by 1870. Wall died in 1871 and his executors ran the enterprise until 1880, whereafter John Wall jr took over. A new brewery was added in 1877, to a design by London-based architects Arthur Kinder.

WALSALL & DISTRICT CLUBS BREWERY LTD, Daw End, Rushall, Walsall.
Founded as the Walsall & District Clubs Co-operative Brewing Society in 1920, it was known as such until 1947. The company had taken over the old Rushall Brewery premises. In 1961 the company, along with Gibbs Mew of Salisbury, bought up the Lancashire Clubs Brewery at Barrowford, forming a new company, Gibbs Keg Breweries Ltd. The Walsall Brewery was sold to Charrington & Co. Ltd that same year.

WALSALL BREWERIES PROPRIETARY LTD, Highgate Brewery, Walsall.
The company was registered in April 1924, with capital of £100, to carry on the business of 'brewers, maltsters etc.' The directors were J.A. Fletcher, Managing Director of Highgate-Walsall Brewery Co. Ltd, and John Lord of the Town Brewery, Short Acre Street, Walsall. See: John Lord.

FREDERICK WARREN, Plough Brewery, Church Street, Brierley Hill.
Founded in 1902. It was offered for auction on 23 March 1926 after the death of its owner. The brewery was not sold, but its seven tied houses realised £13,900.

WEST MIDLANDS WORKING MEN'S CLUB BREWERY CO. LTD, 150 Walsall Road, Willenhall.
Registered in 1920 to acquire Henry Mills Old Oak Brewery. In 1924 the company is entered in the directory as West Midlands Brewery (H. Mills). It closed in 1932.

WHEELWRIGHT'S ARMS BREWERY, Castle Street, Netherton.
Established by Samuel Kendrick Houghton, a wheelwright from Droitwich (hence its name) and opened in 1873. In 1885 the property was leased by Joseph Davies, who succeeded in buying up the freehold in 1888. The brewing side of the business was expanded with the purchase of a number of local beerhouses as outlets. Davies' son, Joseph jr, married Sarah Cooksey, the sister of Solomon Cooksey, brewer at the Old Swan (now more popularly known as Pardoe's). Upon the death of Joseph Davies sr, his grandson, Joseph Henry Davies, succeeded to the business. He sold the pub, brewery and its eleven tied houses to M&B in June 1942. Thereafter brewing ceased at the Wheelwright's Arms.

WILLIAM WHITEHOUSE, Princes End Brewery, 185 Bloomfield Road, Tipton.
Established by Whitehouse in 1862. A brewery and stores were built and in 1880 an office was opened at 103 High Street, Birmingham. The brewery, which closed in 1892, supplied the free trade, including a number of houses in Birmingham.

WOLVERHAMPTON & DUDLEY BREWERIES LTD, Park Brewery, Wolverhampton.
More popularly known as Banks' Brewery, the company was registered in May 1890 to amalgamate Banks & Co., Park Brewery, C.C. Smith's Fox Brewery, Wolverhampton and George Thompson & Sons' Dudley & Victoria Breweries, Dudley, along with 193 tied houses. Edwin Thompson became Managing Director in 1894, and the family were to dominate the firm for over one hundred years. A new sixty-quarter brewery, designed by London-based architect Arthur Kinder, was built in 1898 on the Park Brewery site. In 1909 the company acquired North Worcestershire Breweries Ltd of Stourbridge and Brierley Hill. This was followed by a take-over of the Netherton brewery of John Rolinson & Son Ltd in 1912. The Kidderminster Brewery Co. Ltd was bought up in 1913 with its 126 tied houses. Four years later the City Brewery Co. of Lichfield was purchased, as well as its 200 houses. Banks' closed the brewery, but expanded into the old maltings. In 1934, Dudley brewers Julia Hanson & Sons Ltd was added to the portfolio. Brewing continued here until 1991. Banks' took over Cheshire Inns, along with their fourteen pubs in Cheshire and the Wirral, and bought up five houses in the Manchester area from Wilson's Brewery. The company made two unsuccessful bids to take over Davenport's of Birmingham, in 1983 and 1986. By 1990 W&DB had 800 tied houses. They acquired a further fifty-one houses in the north-east of England in 1992, with the purchase of Cameron's of Hartlepool from Brent Walker. Cameron's Strongarm was introduced to Bank's West Midlands' houses. The company also entered into an agreement with Marston's, permitting the sale of Pedigree in Banks' houses, and Banks' mild in Marston houses. In 1999 the company took over Marston's and the Mansfield Brewery, bringing their stable of tied houses up to 1,763. The company successfully defeated a £453 million hostile takeover bid in August 2000 by the Pubmaster group, but only by a short majority. W&DB sold Cameron's to the neighbouring Castle Eden Brewery and closed down

Above: *Wordsley Brewery beer label.*

Right: *Wordsley Brewery price list, 1897.*

WORDSLEY BREWERY
COMPANY, LTD.,
BREWERS, MALTSTERS,
AND
MINERAL WATER MANUFACTURERS

The large increase in the Company's business during the past twelve months proves beyond a doubt that the Public like a GOOD and PURE Article.

CASK PRICE LIST

MARK.				PRICE PER GAL.
XXXX ...	...	MILD ALE	...	1s. 4d.
XXX ...	...	...	...	1s. 2d.
XX ...	A LIGHT DINNER ALE			1s. 0d.
XT ...	...	...	...	10d.
X ...	...	...	...	8d.
IPA ...	A SPLENDID BITTER ALE			1s. 6d.
PA	SIMILAR TO ABOVE, BUT LIGHER			1s. 4d.
AK ...	...	DITTO		1s. 0d.
IMPERIAL STOUT- SPECIALLY SUITABLE FOR INVALIDS				1s. 6d.
DOUBLE STOUT	...		...	1s. 4d.
PORTER	...		...	1s. 0d.

SPECIAL TERMS TO THE WHOLESALE TRADE

IN 9, 18, 36, & 54 GALLON CASKS.
DELIVERED FREE.
GRAINS ALWAYS FOR SALE

its Mansfield Brewery to appease the Stock Exchange. At the end of 2002, Japanese investment bank, Nomura, the UK's largest pub landlord, made enquiries. W&DB looks vulnerable to a takeover at the right price. Regarding Banks' beers, they are mainly served by an electrical dispenser. Up to the 1970s this took the form of a horizontal glass tube serving a half pint one way, and a full pint if returned. The brewery produces Banks' Mild (ABV 3.5%) and Banks' Bitter (3.8%). They formerly produced Hanson's Mild (3.5%).

WOLVERHAMPTON DISTRICT BREWERY LTD, Victoria Brewery, Wesley Street, Bradley, Wolverhampton.

Registered as a company in December 1898 with a capital of £50,000 to acquire and manage the Bradley Brewery Co. Ltd of Bradley, Staffs., along with its tied houses. The Bradley company was an ale and porter bottling and mineral water business, established in 1885 by Messrs Hudson & Stafford at Wednesbury, with seventeen additional freehold mineral water businesses, including Peck & Kerrison. The Wolverhampton District Brewery went into liquidation in December 1899 and was eventually dissolved on 18 November 1902.

SAMUEL WOODHALL LTD, High Street Brewery, West Bromwich.

A private company founded at 61 High Street, West Bromwich in 1874 and acquiring further premises at 89 High Street soon after. Woodhall's registered as a company on 23 September 1904. The company address changed to 140 High Street in 1921. The brewery closed in 1937, following its takeover by Julia Hanson & Sons.

H. & B. WOODHOUSE, Alma Brewery, Hall Street, Dudley.
Henry and Benjamin Woodhouse originally took over the Alma Inn, formerly the Traveller's Rest, in 1901 and began brewing from there. They moved across the road to the Victoria Brewery when it became available in 1914 and renamed it the Alma Brewery. They moved again in 1917 to the Queen's Cross Brewery, behind the present Lamp Tavern. By then they had fifteen tied houses. The company became Thomas & Benjamin Woodhouse, Queen's Cross Brewery and 7 Castle Street, in 1927. They ceased brewing in 1934.

WORDSLEY BREWERY CO. LTD, Bug Pool Lane (later Brewery Lane), Wordsley.
The brewery was built for Edward Oakes, whose father had kept the Three Furnaces at the Level, Brierley Hill. Oakes moved to Wordsley about 1850 and opened a beerhouse called The Lion. He appears in the trade directories from 1863 as a brewer and maltster, but before that, certainly by 1858, he had established the Lion Brewery at the rear of his public house. In 1875 the County Express (13 February) records that it was producing 'with ease 600 bushels (4,800 gallons of malt) per week'. Oakes is further listed in the directory of 1890 as a 'brewer, maltster, hop merchant, farmer and mineral water manufacturer'. He apparently over-reached himself in his many ventures, and was made bankrupt in 1895. Oakes was obliged to sell the brewery, with its thirteen tied houses, to pay off his creditors. The new owner, George Collis, to whom Oakes had previously mortgaged the brewery, continued brewing as the Wordsley Brewery Co. Ltd. A Mr [George?] Johnson, a professional brewer, was brought in to oversee the operation. Upon his resignation however the company began losing money until, in 1906, facing bankruptcy, the brewery and its thirty-three tied houses were sold to the Hereford & Tredegar Brewery Co. Hereford & Tredegar had no intention of brewing here, and the buildings were sold off to Anthony Bayley, who converted the brewery building itself into a cinema in 1912. The last film was shown in May 1959, and the building was demolished in 1969.

WRIGHT & CO., Smithfield, Wolverhampton.
Common brewers, they are listed as such in the trade directory of 1839. Apparently gone by 1842.

J. & J. YARDLEY & CO. LTD, South Staffordshire Brewery, Market Street, Wolverhampton.
The company also had a London office at 24 Coleman Street, E.C.
The company was registered on 4 November 1897 to acquire as going concerns and to transfer under one management, the business of Messrs Job & Joseph Yardley & Son, brewers and maltsters of Bloxwich and Darlaston, and the South Staffordshire Brewery Co. Ltd at Wolverhampton, along with 134 tied houses. The purchase price was £260,000. In April 1910 the company was taken over by Old Wolverhampton Breweries Ltd.

WHO TOOK OVER WHO

Fig. 1

WOLVERHAMPTON & DUDLEY BREWERIES
(BANKS & CO., – FOX'S BREWERY – GEO.THOMPSON & SON)

North Worcestershire Brewery (1909)	John Rolinson (1912)	Julia Hanson & Sons (1943)
Netherton Old Brewery	Kidderminster Brewery (1913)	Frederick Tandy (1919)
	City Brewery (1917)	Town Brewery (1934)
	John Seedhouse (1955)	Smith & Williams (1934)
	Herbert Newnam (1960)	Samuel Woodhall (1937)
	Cameron's (1992)	Red Lion Brewery (1942)
	Marston's (1999)	

Fig. 2

MITCHELL & BUTLER'S
(HENTRY MITCHELL - WILLIAM BUTLER)

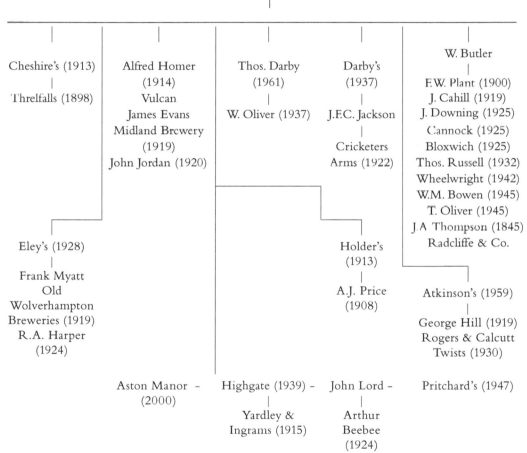

Cheshire's (1913)
|
Threlfalls (1898)

Alfred Homer
(1914)
Vulcan
James Evans
Midland Brewery
(1919)
John Jordan (1920)

Thos. Darby
(1961)
|
W. Oliver (1937)

Darby's
(1937)
|
J.F.C. Jackson
|
Cricketers
Arms (1922)

W. Butler
|
F.W. Plant (1900)
J. Cahill (1919)
J. Downing (1925)
Cannock (1925)
Bloxwich (1925)
Thos. Russell (1932)
Wheelwright (1942)
W.M. Bowen (1945)
T. Oliver (1945)
J.A Thompson (1845)
Radcliffe & Co.

Eley's (1928)
|
Frank Myatt
Old
Wolverhampton
Breweries (1919)
R.A. Harper
(1924)

Holder's
(1913)
|
A.J. Price
(1908)

Atkinson's (1959)
|
George Hill (1919)
Rogers & Calcutt
Twists (1930)

Aston Manor –
(2000)

Highgate (1939) –
|
Yardley &
Ingrams (1915)

John Lord –
|
Arthur
Beebee
(1924)

Pritchard's (1947)

Fig. 3

ANSELL'S

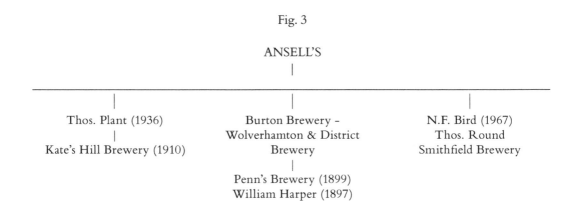

Thos. Plant (1936)
|
Kate's Hill Brewery (1910)

Burton Brewery –
Wolverhamton & District
Brewery
|
Penn's Brewery (1899)
William Harper (1897)

N.F. Bird (1967)
Thos. Round
Smithfield Brewery

BIBLIOGRAPHY

Barber, Norman, *A Century of British Brewing, 1890-1990*. Brewing History Society, 1994.

Billingham, Clive, *Home Brewing in the Black Country*. Blackcountryman, Spring 1995, Vol. 28, No. 2.
Brewers' Journal, 1879-1960.

Cox, David, The Wordsley Brewery. *The Journal of the Brewery History Society*. No. 94, Winter 1998.

Dunn, Mike, *Local Brew*. Robert Hale, 1986.

Haden, H. Jack, *Wordsley 'Lymp' and its Proprietor*. Blackcountryman, Spring 1984, Vol. 17 No.2 and Summer 1984, Vol. 17, No. 3.

Histories of Birmingham Companies, No. 10. *Mitchell & Butler's*, Birmingham Sketch, 1958.

Kelly's Directories of Staffordshire and Worcestershire, 1860-1940.

Lloyd, Keith J., *The Highgate Brewery, Walsall*. Black Country Society, 1983.

Manuel of British & Foreign Brewery Companies (later the Brewery Manual), 1899-1996.

Parliamentary Papers 1890-91 (c.28) LXVIII. Licensed Properties.
– 1895 (C.96) LXXXVIII.

Peaty, Ian P., *You Brew Good Ale*. Sutton Publishing, Stroud, 1997.

Richards, John, *The Pubs & Breweries of the Old Dudley Borough*. Real Ale Books, Dudley, 1989.
– 'The Wordsley Brewery Company Limited'. Brewery History No. 99, Spring 2000.
– 'Thomas Booth – Black Country Brewer'. Journal of the Brewery History Society, No. 83, Spring 1996.
– *A History of Holden's Black Country Brewery*. Manchester, 1986.
– *The History of Batham's Black Country Brewers*. Real Ale Books, 1993

Somers, F. & K.M., *Halas Hales Halesowen*. 1932.

Stevens, Geoff, *The Brewers Swan Song. Around the Black Country*. Black Country Society, 1973.

INDEX

N.B. Retail Brewers have not been indexed as they appear in alphabetical order within the text by town or village.

XYZ

Other local titles published by The History Press

Dudley
DAVID CLARE

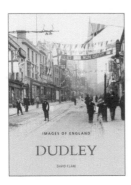

This fascinating collection highlights some of the changes and developments that have taken place in Dudley over the last 150 years. Originally a medieval market town, Dudley was shaped by the Industrial Revolution. Every aspect of Dudley is explored, from the people and buildings of this strong community to market days, transport and shops; from the highest tower of the castle to the subterranean limestone caverns beneath Castle Hill.

0 7524 3534 5

Wolverhampton Pubs
ALEC BREW

This comprehensive volume of archive images recalls the intriguing history of many of Wolverhampton's pubs, from the First World War through to the present day. Illustrated with over 200 pictures, Alec Brew charts the changing façades of the town's inns, from street corner pubs to the imposing new pubs built between the wars, and offers an insight into the popularity and changing role of Wolverhampton's pubs.

0 7524 3156 0

Gloucestershire Pubs and Breweries
TIM EDGELL AND GEOFF SANDLES

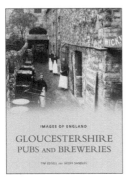

Illustrated with over 200 old photographs, postcards and promotional advertisements, this absorbing collection offers the reader an insight into Gloucestershire's pubs and breweries past and present. Included are images of the Cheltenham Original Brewery, Nailsworth Brewery and the Stroud Brewery Company, as well as many snapshots of local pubs and landlords. This book is sure to appeal to anyone interested in the history of the county's brewing industry.

0 7524 3524 8

Walsall Leather Industry The World's Saddlers
MICHAEL GLASSON

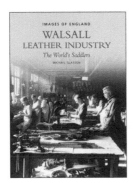

For nearly 200 years Walsall has been a major centre of leather industry, exporting saddles, bridles and a variety of horse equipment to most corners of the world. At its peak the industry employed over 10,000 men and women, with the British Army being the single biggest customer. These days Walsall maintains an international reputation for it's products, and not surprisingly the town has been called the saddlery 'capital' of the world.

0 7524 2793 8

If you are interested in purchasing other books published by The History Press, or in case you have difficulty finding any of our books in your local bookshop, you can also place orders directly through our website

www.thehistorypress.co.uk